Garrett Putman Serviss

Astronomy With An Opera-Glass

Garrett Putman Serviss

Astronomy With An Opera-Glass

ISBN/EAN: 9783337030032

Printed in Europe, USA, Canada, Australia, Japan

Cover: Foto ©berggeist007 / pixelio.de

More available books at **www.hansebooks.com**

ASTRONOMY

WITH AN OPERA-GLASS

A POPULAR INTRODUCTION TO THE
STUDY OF THE STARRY HEAVENS WITH THE
SIMPLEST OF OPTICAL INSTRUMENTS

WITH MAPS AND DIRECTIONS TO FACILITATE THE RECOGNITION
OF THE CONSTELLATIONS AND THE PRINCIPAL STARS
VISIBLE TO THE NAKED EYE

BY

GARRETT P. SERVISS

" Known are their laws ; in harmony unroll
The nineteen-orbed cycles of the Moon.
And all the signs through which Night whirls her car
From belted Orion back to Orion and his dauntless Hound,
And all Poseidon's, all high Zeus' stars
Bear on their beams true messages to man."
POSTE'S ARATUS.

THIRD EDITION

NEW YORK
D. APPLETON AND COMPANY
LONDON: CAXTON HOUSE, PATERNOSTER SQUARE
1890

TO THE READER.

In the pages that follow, the author has endeavored to encourage the study of the heavenly bodies by pointing out some of the interesting and marvelous phenomena of the universe that are visible with little or no assistance from optical instruments, and indicating means of becoming acquainted with the constellations and the planets. Knowing that an opera-glass is capable of revealing some of the most beautiful sights in the starry dome, and believing that many persons would be glad to learn the fact, he set to work with such an instrument and surveyed all the constellations visible in the latitude of New York, carefully noting everything that it seemed might interest amateur star-gazers. All the objects thus observed have not been included in this book, lest the multiplicity of details should deter or discourage the very readers for whom it was specially written. On the other hand, there is nothing described as visible with an opera-glass or a field-glass which the author has not seen with an instrument of that description, and which any person possessing eyesight of average quality and a competent glass should not be able to discern.

But, in order to lend due interest to the subject, and place it before the reader in a proper light and true perspective, many facts have been stated concerning the objects described, the ascertainment of which has required the aid of powerful telescopes, and to observers with such instruments is reserved the noble pleasure of confirming with their own eyes those

wonderful discoveries which the looker with an opera-glass
can not hope to behold unless, happily, he should be spurred
on to the possession of a telescope. Yet even to glimpse dimly
these distant wonders, knowing what a closer view would reveal, is a source of no mean satisfaction, while the celestial
phenomena that lie easily within reach of an opera-glass are
sufficient to furnish delight and instruction for many an
evening.

It should be said that the division of the stars used in this
book into the "Stars of Spring," "Stars of Summer," "Stars
of Autumn," and "Stars of Winter," is purely arbitrary, and
intended only to indicate the seasons when certain constellations are best situated for observation or most conspicuous.

The greater part of the matter composing this volume appeared originally in a series of articles contributed by the author to "The Popular Science Monthly" in 1887–'88. The
reception that those articles met with encouraged him to re-
vise and enlarge them for publication in the more permanent
form of a book.

G. P. S.

BROOKLYN, N. Y., *September, 1888.*

CONTENTS.

ASTRONOMY WITH AN OPERA-GLASS.

INTRODUCTION.

STAR-GAZING was never more popular than it is now. In every civilized country many excellent telescopes are owned and used, often to very good purpose, by persons who are not practical astronomers, but who wish to see for themselves the marvels of the sky, and who occasionally stumble upon something that is new even to professional star-gazers. Yet, notwithstanding this activity in the cultivation of astronomical studies, it is probably safe to assert that hardly one person in a hundred knows the chief stars by name, or can even recognize the principal constellations, much less distinguish the planets from the fixed stars. And of course they know nothing of the intellectual pleasure that accompanies a knowledge of the stars. Modern astronomy is so rapidly and wonderfully linking the earth and the sun together, with all the orbs of space, in the bonds of close physical relationship, that a person of education and general intelligence can offer no valid excuse for not knowing where to look for Sirius or Aldebaran, or the Orion nebula, or the planet Jupiter. As Australia and New Zealand and the islands of the sea are made a part of the civilized world through the expanding influence of commerce and cultivation, so the suns and planets around us are, in a certain sense, falling under the dominion of the restless and resistless mind of man. We have come to possess vested intellectual interests in Mars and Saturn, and in the sun and all his multitude of fellows, which nobody can afford to ignore.

A singular proof of popular ignorance of the starry heavens, as well as of popular curiosity concerning any uncommon celestial phenomenon, is furnished by the curious notions prevailing about the planet Venus. When Venus began to attract general attention in the western sky in the early evenings of the spring of 1887, speculation quickly became rife about it, particularly on the great Brooklyn Bridge. As the planet hung dazzlingly bright over the New Jersey horizon, some people appeared to think it was the light of Liberty's torch, mistaking the bronze goddess's real flambeau for a part of the electric-light system of the metropolis. Finally (to judge from the letters written to the newspapers, and the questions asked of individuals supposed to know something about the secrets of the sky), the conviction seems to have become pretty widely distributed that the strange light in the west was no less than an electrically illuminated balloon, nightly sent skyward by Mr. Edison, for no other conceivable reason than a wizardly desire to mystify his fellow-men. I have positive information that this ridiculous notion has been actually entertained by more than one person of intelligence. And as Venus glowed with increasing splendor in the serene evenings of June, she continued to be mistaken for some petty artificial light instead of the magnificent world that she was, sparkling out there in the sunshine like a globe of burnished silver. Yet Venus as an evening star is not so rare a phenomenon that people of intelligence should be surprised at it. Once in every 584 days she reappears at the same place in the sunset sky—

> "Gem of the crimson-colored even,
> Companion of retiring day."

No eye can fail to note her, and as the nearest and most beautiful of the Earth's sisters it would seem that everybody should be as familiar with her appearance as with the face of a friend. But the popular ignorance of Venus, and the other members of the planetary family to which our mother, the Earth, belongs, is only an index of the denser ignorance concerning the stars—the brothers of our great father, the Sun. I believe this ignorance is largely due to mere indifference, which, in its turn, arises from a false and pedantic method of presenting astronomy as a creature of mathematical formulæ, and a humble handmaiden of the art of navigation. I do not, of course, mean to cast doubt upon the scientific value of technical work in astronomy. The science could not exist without it. Those who have made the spectroscope reveal the composition of the sun and stars, and who are now making photography picture the heavens as they are, and even reveal phenomena which lie beyond the range of human vision, are the men who have taken astronomy out of its swaddling-clothes, and set it on its feet as a progressive science. But when one sees the depressing and repellent effect that has evidently been produced upon the popular mind by the ordinary methods of presenting astronomy, one can not resist the temptation to utter a vigorous protest, and to declare that this glorious science is not the grinning mathematical skeleton that it has been represented to be.

Perhaps one reason why the average educated man or woman knows so little of the starry heavens is because it is popularly supposed that only the most powerful telescopes and costly instruments of the observatory are capable of dealing with them. No greater mistake could be made. It does not require an optical instrument of any kind, nor much labor, as compared with that expended in the acquirement of some polished accomplishments regarded as indispensable, to give one an acquaintance with the stars and planets which will be not only pleasurable but useful. And with the aid of an opera-glass most interesting, gratifying, and, in some instances, scientifically valuable observations may be made in the heavens. I have more than once heard persons who knew nothing about the stars, and probably cared less, utter exclamations of surprise and delight when persuaded to look at certain parts of the sky with a good glass, and thereafter manifest an interest in astronomy of which they would formerly have believed themselves incapable.

Being convinced that whoever will survey the heavens with a good opera-glass will feel repaid many fold for his time and labor, I have undertaken to point out some of the objects most worthy of attention, and some of the means of making acquaintance with the stars.

First, a word about the instrument to be used. Galileo made his famous discoveries with what was, in principle of construction, simply an opera-glass. This form of telescope was afterward abandoned because very high magnifying powers could not be employed with it, and the field of view was restricted. But, on account of its brilliant illumination of objects looked at, and its convenience of form, the opera-glass is still a valuable and, in some respects, unrivaled instrument of observation.

In choosing an opera-glass, see first that the object-glasses are achromatic, although this caution is hardly necessary, for all modern opera-glasses, worthy of the name, are made with achromatic objectives. But there are great differences in the quality of the work. If a glass shows a colored fringe around a bright object, reject it. Let the diameter of the object-glasses, which are the large lenses in the end farthest from the eye, be not less than an inch and a half. The magnifying power should be at least three or four diameters. A familiar way of estimating the magnifying power is by looking at a brick wall through one barrel of the opera-glass with one eye, while the other eye sees the wall without the intervention of the glass. Then notice how many bricks seen by the naked eye are required to equal in thickness one brick seen through the glass. That number represents the magnifying power.

The instrument used by the writer in making most of the observations for this book has object-glasses 1·6 inch in diameter, and a magnifying power of about 3·6 times.

See that the fields of view given by the two barrels of the opera-glass coincide, or blend perfectly together. If one appears to partially overlap the other when looking at a distant object, the effect is very annoying. This fault arises from the barrels of the opera-glass being placed too far apart, so that their optical centers do not coincide with the centers of the observer's eyes.

Occasionally, on account of faulty centering of the lenses, a double image is given of objects looked at, as illustrated in the accompanying cut. In such a case the glass is worthless; but if the effect is simply the addition of a small, crescent-shaped extension on one side of the field of view without any reduplication, the fault may be overlooked, though it is far better to select a glass that gives a perfectly round field. Some glasses have an arrangement for adjusting the distance between the barrels to suit the eyes of different persons, and it would be well if all were made adjustable in the same way.

A VERY BAD FIELD.

Don't buy a cheap glass, but don't waste your money on fancy mountings. What the Rev. T. W. Webb says of telescopes is equally true of opera-glasses: "Inferior articles may be showily got up, and the outside must go for nothing." There are a few makers whose names, stamped upon the instrument, may generally be regarded as a guarantee of excellence. But the best test is that of actual performance. I have a field-glass which I found in a pawn-shop, that has no maker's name upon it, but in some respects is quite capable of bearing comparison with the work of the best advertised opticians. And this leads me to say that, by the exercise of good judgment, one may occasionally purchase superior glasses at very reasonable prices in the pawn-shops. Ask to be shown the old and well-tried articles ; you may find among them a second-hand glass of fine optical properties. If the lenses are not injured, one need not trouble one's self about the worn appearance of the outside of the instrument ; so much the more evidence that somebody has found it well worth using.

A good field or marine glass is in some respects better than an opera-glass for celestial observations. It possesses a much higher magnifying power, and this gives sometimes a decided advantage. But, on the other hand, its field of view is smaller, rendering it more difficult to find and hold objects. Besides, it does not present as brilliant views of scattered star-clusters as an opera-glass does. For the benefit of those who possess field-glasses, however, I have included in this brief survey certain objects that lie just beyond the reach of opera-glasses, but can be seen with the larger instruments.

I have thought it advisable in the descriptions of the constellations which follow to give some account of their mythological origin, both because of the historical interest which attaches to it, and because, while astronomers have long since banished the constellation figures from their maps, the names which the constellations continue to bear require some explanation, and they possess a literary and romantic interest which can not be altogether disregarded in a work that is not intended for purely scientific readers.

CHAPTER I.

HAVING selected your glass, the next thing is to find the stars. Of course, one could sweep over the heavens at random on a starry night and see many interesting things, but he would soon tire of such aimless occupation. The observer must know what he is looking at in order to derive any real pleasure or satisfaction from the sight.

It really makes no difference at what time of the year such observations are begun, but for convenience I will suppose that they are begun in the spring. We can then follow the revolution of the heavens through a year, at the end of which the diligent observer will have acquired a competent knowledge of the constellations. The circular map, No. 1, represents the appearance of the heavens at midnight on the 1st of March, at eleven o'clock on the 15th of March, at ten o'clock on the 1st of April, at nine o'clock on the 15th of April, and at eight o'clock on the 1st of May. The reason why a single map can thus be made to show the places of the stars at different hours in different months will be plain upon a little reflection. In consequence of the earth's annual journey around the sun, the whole heavens make one apparent revolution in a year. This revolution, it is clear, must be at the rate of 30° in a month, since the complete circuit comprises 360°. But, in addition to the annual revolution, there is a diurnal revolution of the heavens which is caused by the earth's daily rotation upon its axis, and this revolution must, for a similar reason, be performed at the rate of 15° for each

of the twenty-four hours. It follows that in two hours of the
daily revolution the stars will change their places to the same

MAP 1.

extent as in one month of the annual revolution.   It follows
also that, if one could watch the heavens throughout the
whole twenty-four hours, and not be interrupted by daylight,
he would behold the complete circuit of the stars just as he
would do if, for a year, he should look at the heavens at a
particular hour every night.   Suppose that at nine o'clock on

the 1st of June we see the star Spica on the meridian; in consequence of the rotation of the earth, two hours later, or at eleven o'clock, Spica will be 30° west of the meridian. But that is just the position which Spica would occupy at nine o'clock on the 1st of July, for in one month (supposing a month to be accurately the twelfth part of a year) the stars shift their places 30° toward the west. If, then, we should make a map of the stars for nine o'clock on the 1st of July, it would answer just as well for eleven o'clock on the 1st of June, or for seven o'clock on the 1st of August.

The center of the map is the zenith, or point overhead. The reader must now exercise his imagination a little, for it is impossible to represent the true appearance of the concave of the heavens on flat paper. Holding the map over your head, with the points marked East, West, North, and South in their proper places, conceive of it as shaped like the inside of an open umbrella, the edge all around extending clear down to the horizon. Suppose you are facing the south, then you will see, up near the zenith, the constellation of Leo, which can be readily recognized on the map by six stars that mark out the figure of a sickle standing upright on its handle. The large star in the bottom of the handle is Regulus. Having fixed the appearance and situation of this constellation in your mind, go out-of-doors, face the south, and try to find the constellation in the sky. With a little application you will be sure to succeed.

Using Leo as a basis of operations, your conquest of the sky will now proceed more rapidly. By reference to the map you will be able to recognize the twin stars of Gemini, southwest of the zenith and high up; the brilliant lone star, Procyon, south of Gemini; the dazzling Sirius, flashing low down in the southwest; Orion, with all his brilliants, blazing in the west; red Aldebaran and the Pleiades off to his right; and Capella, bright as a diamond, high up above Orion, toward the north. In the southeast you will recognize the quadrilateral of Corvus, with the remarkably white star Spica glittering east of it.

Next face the north. If you are not just sure where north is, try a pocket-compass. This advice is by no means unnecessary, for there are many intelligent persons who are unable to indicate true north within many degrees, though standing on their own doorstep. Having found the north point as near as you can, look upward about forty degrees from the horizon, and you will see the lone twinkler called the north or pole star. Forty degrees is a little less than half-way from the horizon to the zenith.

By the aid of the map, again, you will be able to find, high up in the northeast, near the zenith, the large dipper-shaped figure in Ursa Major, and, when you have once noticed that the two stars in the outer edge of the bowl of the Dipper point almost directly to the pole-star, you will have an unfailing means of picking out the latter star hereafter, when in doubt.* Continuing the curve of the Dipper-handle, in the northeast, your eye will be led to a bright reddish star, which is Arcturus, in the constellation Boötes.

In the same way you will be able to find the constellations Cassiopeia, Cepheus, Draco, and Perseus. Don't expect to accomplish it all in an hour. You may have to devote two or three evenings to such observation, and make many trips indoors to consult the map, before you have mastered the subject; but when you have done it you will feel amply repaid for your exertions, and you will have made for yourself silent friends in the heavens that will beam kindly upon you, like old neighbors, on whatever side of the world you may wander.

Having fixed the general outlines and location of the constellations in your mind, and learned to recognize the chief stars, take your opera-glass and begin with the constellation

---

* Let the reader remember that the distance between the two stars in the brim of the bowl of the Dipper is about ten degrees, and he will have a measuring-stick that he can apply in estimating other distances in the heavens.

Leo and the star Regulus. Contrive to have some convenient rest for your arms in holding the glass, and thus obtain not only comfort but steadiness of vision. A lazy-back chair makes a capital observing-seat. Be very particular, too, to get a sharp focus. Remember that no two persons' eyes are alike, and that even the eyes of the same observer occasionally require a change. In looking for a difficult object, I have sometimes suddenly brought the sought-for phenomenon into view by a slight turn of the focusing-screw.

You will at once be gratified by the increased brilliancy of the star as seen by the glass. If the night is clear, it will glow like a diamond. Yet Regulus, although ranked as a first-magnitude star, and of great repute among the ancient astrologers, is far inferior in brilliancy to such stars as Capella and Arcturus, to say nothing of Sirius.

By consulting map No. 2 you will next be able to find the celebrated star bearing the name of the Greek letter Gamma (γ). If you had a telescope, you would see this star as a close and beautiful double, of contrasted colors. But it is optically double, even with an opera-glass. You can not fail to see a small star near it, looking quite close if the magnifying power of your glass is less than three times. You will be struck by the surprising change of color in turning from Regulus to Gamma—the former is white and the latter deep yellow. It will be well to look first at one and then at the other, several times, for this is a good instance of what you will meet with many times in your future surveys of the heavens—a striking contrast of color in neighboring stars. One can thus comprehend that there is more than one sense in which to understand the Scriptural declaration that "one star differeth from another in glory." The radiant point of the famous November meteors, which, in 1833 and 1866, filled the sky with fiery showers, is near Gamma. Turn next to the star in Leo marked Zeta (ζ). If your glass is a pretty large and good one, and your eye keen, you will easily see three

2

minute companion stars keeping company with Zeta, two on
the southeast, and one, much closer, toward the north.   The

MAP 2.

nearest of the two on the south is faint, being only between
the eighth and ninth magnitude, and will probably severely
test your powers of vision.   Next look at Epsilon (ε), and
you will find near it two seventh-magnitude companions,
making a beautiful little triangle.

Away at the eastern end of the constellation, in the tail
of the imaginary Lion, upon whose breast shines Regulus, is
the star Beta (β) Leonis, also called Denebola.   It is almost
as bright as its leader, Regulus, and you will probably be
able to catch a tinge of blue in its rays.   South of Denebola,
at a distance of nineteen minutes of arc, or somewhat more
than half the apparent diameter of the moon, you will see a
little star of the sixth magnitude, which is one of the several

"companions" for which Denebola is celebrated. There is another star of the eighth magnitude in the same direction from Denebola, but at a distance of less than five minutes, and this you may be able to glimpse with a powerful field-glass, under favorable conditions. I have seen it well with a field-glass of 1·6-inch aperture, and a magnifying power of seven times. But it requires an experienced eye and steady vision to catch this shy twinkler.

When looking for a faint and difficult object, the plan pursued by telescopists is to avert the eye from the precise point upon which the attention is fixed, in order to bring a more sensitive part of the retina into play than that usually employed. Look toward the edge of the field of view, while the object you are seeking is in the center, and then, if it can be seen at all with your glass, you will catch sight of it, as it were, out of the corner of your eye. The effect of seeing a faint star in this way, in the neighborhood of a large one, whose rays hide it from direct vision, is sometimes very amusing. The little star seems to dart out into view as through a curtain, perfectly distinct, though as immeasurably minute as the point of a needle. But the instant you direct your eyes straight at it, presto! it is gone. And so it will dodge in and out of sight as often as you turn your eyes.

If you will sweep carefully over the whole extent of Leo, whose chief stars are marked with their Greek-letter names on our little map, you will be impressed with the power of your glass to bring into sight many faint stars in regions that seem barren to the naked eye. An opera-glass of 1·5 aperture will show ten times as many stars as the naked eye can see.

A word about the "Lion" which this constellation is supposed to represent. It requires a vivid imagination to perceive the outlines of the celestial king of beasts among the stars, and yet somebody taught the people of ancient India and the old Egyptians to see him there, and there he has remained since the dawn of history. Modern astronomers strike him out of their charts, together with all the picturesque multitude of beasts and birds and men and women that bear him company, but they can not altogether banish him, or any of his congeners, for the old names, and, practically, the old outlines of the constellations are retained, and always will be retained. The Lion is the most conspicuous figure in the celebrated zodiac of Dendera; and, indeed, there is evidence that before the story of Hercules and his labors was told this lion was already imagined shining among the stars. It was characteristic of the Greeks that they seized him for their own, and tried to rob him of his real antiquity by pretending that Jupiter had placed him among the stars in commemoration of Hercules's victory over the Nemæan lion. In the Hebrew zodiac Leo represented the Lion of Judah. It was thus always a lion that the ancients thought they saw in this constellation.

In the old star-maps the Lion is represented as in the act of springing upon his prey. His face is to the west, and the star Regulus is in his heart. The sickle-shaped figure covers his breast and head, Gamma being in the shoulder, Zeta in the mane of the neck, Mu and Epsilon in the cheek, and Lambda in the jaws. The fore-paws are drawn up to the breast and represented by the stars Zi and Omicron. Denebola is in the tuft of the tail. The hind-legs are extended downward at full length, in the act of springing. Starting from the star Delta in the hip, the row consisting of Theta, Iota, Tau, and Upsilon, shows the line of the hind-legs.

Leo had an unsavory reputation among the ancients because of his supposed influence upon the weather. The greatest heat of summer was felt when the sun was in this constellation :

> " Most scorching is the chariot of the Sun,
> And waving spikes no longer hide the furrows
> When he begins to travel with the Lion."

Looking now westwardly from the Sickle of Leo, at a distance about equal to twice the length of the Sickle, your eye will be caught by a small silvery spot in the sky lying nearly between two rather faint stars. This is the famous Præsepe, or Manger, in the center of the constellation Cancer. The two stars on either side of it are called the Aselli, or the Ass's Colts, and the imagination of the ancients pictured them feeding from their silver manger. Turn your glass upon the Manger and you will see that it consists of a crowd of little stars, so small and numerous that you will probably not undertake to count them, unless you are using a large field-glass. Galileo has left a delightful description of his surprise and gratification when he aimed his telescope at this curious cluster and other similar aggregations of stars and discovered what they really were. Using his best instrument, he was able to count thirty-six stars in the Manger. The Manger was a famous weather-sign in olden times, and Aratus, in his "Diosemia," advises his readers to—

". . . watch the Manger: like a little mist
Far north in Cancer's territory it floats.
Its confines are two faintly glimmering stars;
These are two asses that a manger parts,
Which suddenly, when all the sky is clear,
Sometimes quite vanishes, and the two stars
Seem to have closer moved their sundered orbs.
No feeble tempest then will soak the leas;
A murky manger with both stars
Shining unaltered is a sign of rain."

Like other old weather-saws, this probably possesses a gleam of sense, for it is only when the atmosphere is perfectly transparent that the Manger can be clearly seen; when the air is thick with mist, the harbinger of coming storm, it fades from sight.

The constellation Cancer, or the Crab, was represented by the Egyptians under the figure of a scarabæus. The observer will probably think that it is as easy to see a beetle as a crab there. Cancer, like Leo, is one of the twelve constellations of the Zodiac, the name applied to the imaginary zone 16° degrees wide and extending completely around the heavens, the center of which is the ecliptic or annual path of the sun. The names of these zodiacal constellations, in their order, beginning at the west and counting round the circle, are: Aries, Taurus, Gemini, Cancer, Leo, Virgo, Libra, Scorpio, Sagittarius, Capricornus, Aquarius, and Pisces. Cancer has given its name to the circle called the Tropic of Cancer, which indicates the greatest northerly declination of the sun in summer, and which he attains on the 21st or 22d of June. But, in consequence of the precession of the equinoxes, all of the zodiacal constellations are continually shifting toward the east, and Cancer has passed away from the place of the summer solstice, which is now to be found in Gemini.

Below the Manger, a little way toward the south, your eye will be caught by a group of four or five stars of about the same brightness as the Aselli. This marks the head of Hydra, and the glass will show a striking and beautiful geometrical arrangement of the stars composing it. Hydra is a very long constellation, and trending southward and eastward from the head it passes underneath Leo, and, sweeping pretty close down to the horizon, winds away under Corvus, the tail reaching to the eastern horizon. The length of this skyey serpent is about 100°. Its stars are all faint, except Alphard, or the Hydra's Heart, a second-magnitude star, remarkable for its lonely situation, southwest of Regulus. A line from Gamma Leonis through Regulus points it out. It is worth looking at with the glass on account of its rich orange-tint.

Hydra is fabled to be the hundred-headed monster that was slain by Hercules. It must be confessed that there is nothing very monstrous about it now except its length. The most timid can look upon it without suspecting its grisly origin.

Coming back to the Manger as a starting-point, look well up to the north and west, and at a distance somewhat less than that between Regulus and the Manger you will see a pair of first-magnitude stars, which you will hardly need to be informed are the celebrated Twins, from which the constellation Gemini takes its name. The star marked *a* in the map is Castor, and the star marked *β* is Pollux. No classical reader needs to be reminded of the romantic origin of these names.

A sharp contrast in the color of Castor and Pollux comes out as soon as the glass is turned upon them. Castor is white, with occasionally, perhaps, a suspicion of a green ray in its light. Pollux is deep yellow. Castor is a celebrated double star, but its components are far too close to be separated with an opera-glass, or even the most powerful field-glass. You will be at once interested by the singular *cortége* of small stars by which both Castor and Pollux are surrounded. These little attendant stars, for such they seem, are arrayed in symmetrical groups — pairs, triangles, and other figures—which, it seems difficult to believe, could be unintentional, although it would be still more difficult to suggest any reason why they should be arranged in that way.

Our map will show you the position of the principal stars of the constellation. Castor and Pollux are in the heads of the Twins, while the row of stars shown in the map Xi (*ξ*), Gamma (*γ*), Nu (*ν*), Mu (*μ*), and Eta (*η*), marks their feet, which are dipped in the edge of the Milky-Way. One can spend a profitable and pleasurable half-hour in exploring the wonders of Gemini. The whole constellation, from head to foot, is gemmed with stars which escape the naked eye, but it sparkles like a bead-spangled garment when viewed with the glass. · Owing to the presence of the Milky-Way, the spectacle around the feet of the Twins is particularly magnificent. And here the possessor of a good opera-glass can get a fine view of a celebrated star-cluster known in the

catalogues as 35 M. It is situated a little distance northwest
of the star Eta, and is visible to the naked eye, on a clear,

MAP 3.

moonless night, as a nebulous speck. With a good glass you
will see two wonderful streams of little stars starting, one
from Eta and the other from Mu, and running parallel toward
the northwest ; 35 M is situated between these star-streams.
The stars in the cluster are so closely aggregated that you
will be able to clearly separate only the outlying ones. The
general aspect is like that of a piece of frosted silver over
which a twinkling light is playing. A field-glass brings out
more of the component stars. The splendor of this starry
congregation, viewed with a powerful telescope, may be
guessed at from Admiral Smyth's picturesque description :
"It presents a gorgeous field of stars, from the ninth to the
sixteenth magnitude, but with the center of the mass less
rich than the rest. From the small stars being inclined to
form curves of three or four, and often with a large one at
the root of the curve, it somewhat reminds one of the bursting of a sky-rocket." And Webb adds that there is an

"elegant festoon near the center, starting with a reddish star."

No one can gaze upon this marvelous phenomenon, even with the comparatively low powers of an opera-glass, and reflect that all these swarming dots of light are really suns, without a stunning sense of the immensity of the material universe.

It is an interesting fact that the summer solstice, or the point which the sun occupies when it attains its greatest northerly declination, on the longest day of the year, is close by this great cluster in Gemini. In the glare of the sunshine those swarming stars are then concealed from our sight, but with the mind's eye we can look past and beyond our sun, across the incomprehensible chasm of space, and behold them still shining, their commingled rays making our great God of Day seem but a lonely wanderer in the expanse of the universe.

It was only a short distance southwest of this cluster that one of the most celebrated discoveries in astronomy was made. There, on the evening of March 13, 1781, William Herschel observed a star whose singular aspect led him to put a higher magnifying power on his telescope. The higher power showed that the object was not a star but a planet, or a comet, as Herschel at first supposed. It was the planet Uranus, whose discovery "at one stroke doubled the breadth of the sun's dominions."

The constellation of Gemini, as the names of its two chief stars indicate, had its origin in the classic story of the twin sons of Jupiter and Leda :

"Fair Leda's twins, in time to stars decreed,
    One fought on foot, one curbed the fiery steed."

Castor and Pollux were regarded by both the Greeks and the Romans as the patrons of navigation, and this fact crops out very curiously in the adventures of St. Paul. After his disastrous shipwreck on the island of Melita he embarked again on a more prosperous voyage in a ship bearing the name of these very brothers. "And after three months," writes the celebrated apostle (Acts xxviii, 11) "we departed in a ship of Alexandria, which had wintered in the isle, whose sign was Castor and Pollux." We may be certain that Paul was acquainted with the constellation of Gemini, not only because he was skilled in the learning of his times, but because, in his speech on Mars Hill, he quoted a line from the opening stanzas of Aratus's "Phenomena," a poem in which the constellations are described.

The map will enable you next to find Procyon, or the Little Dog-Star, more than twenty degrees south of Castor and Pollux, and almost directly below the Manger. This star will interest you by its golden-yellow color and its brightness, although it is far inferior in the latter respect to Sirius, or the Great Dog-Star, which you will see flashing splendidly far down beneath Procyon in the southwest. About four degrees northwest of Procyon is a third-magnitude star, called Gomelza, and the glass will show you two small stars which make a right-angled triangle with it, the nearer one being remarkable for its ruddy color.

Procyon is especially interesting because it is attended by an invisible star, which, while it has escaped all efforts to detect it with powerful telescopes, nevertheless reveals its presence by the effect of its attraction upon Procyon. It is a curious fact that both of the so-called Dog-Stars are thus attended by obscure or dusky companion-stars, which, notwithstanding their lack of luminosity, are of great magnitude. In the case of Sirius, the improvement in telescopes has brought the mysterious attendant into view, but Procyon's mate remains hidden from our eyes. But it can not escape the ken of the mathematician, whose penetrating mental vision has, in more than one instance, outstripped the discoveries of the telescope. Almost half a century ago the famous Bessel announced his conclusion—in the light of later

developments it may well be called discovery—that both Sirius and Procyon were binary systems, consisting each of a
visible and an invisible star. He calculated the probable
period of revolution, and found it to be, in each case, approximately fifty years. Sixteen years after Bessel's death,
one of Alvan Clark's unrivaled telescopes at last revealed the
strange companion of Sirius, a huge body, half as massive as
the giant Dog-Star itself, but ten thousand times less brilliant, and more recent observations have shown that its period of revolution is within six or seven months of the fifty
years assigned by Bessel. If some of the enormous telescopes that have been constructed in the past few years
should succeed in rendering Procyon's companion visible also,
it is highly probable that Bessel's prediction would receive
another substantial fulfillment.

The mythological history of Canis Minor is somewhat obscure. According to various accounts it represents one of
Diana's hunting-dogs, one of Orion's hounds, the Egyptian
dog-headed god Anubis, and one of the dogs that devoured
their master Actæon after Diana had turned him into a stag.
The mystical Dr. Seiss leaves all the ancient myth-makers
far in the rear, and advances a very curious theory of his
own about this constellation, in his "Gospel in the Stars,"
which is worth quoting as an example of the grotesque
fancies that even in our day sometimes possess the minds
of men when they venture beyond the safe confines of this
terraqueous globe. After summarizing the various myths
we have mentioned, he proceeds to identify Procyon, put-
ting the name of the chief star for the constellation, "as
the starry symbol of those heavenly armies which came forth
along with the King of kings and Lord of lords to the battle
of the great day of God Almighty, to make an end of mis-
rule and usurpation on earth, and clear it of all the wild
beasts which have been devastating it for these many ages."
The reader will wonder all the more at this rhapsody

after he has succeeded in picking out the modest Little
Dog in the sky.

Sirius, Orion, Aldebaran, and the Pleiades, all of which
you will perceive in the west and southwest, are generally
too much involved in the mists of the horizon to be seen to
the best advantage at this season, although it will pay you
to take a look through the glass at Sirius. But the splendid
star Capella, in the constellation Auriga, may claim a moment's attention. You will find it high up in the northwest,
half-way between Orion and the pole-star, and to the right
of the Twins. It has no rival near, and its creamy-white
light makes it one of the most beautiful as well as one of
the most brilliant stars in the heavens. Its constitution, as
revealed by the spectroscope, resembles that of our sun, but
the sun would make but a sorry figure if removed to the side
of this giant star. About seven and a half degrees above
Capella, and a little to the left, you will see a second-magnitude star called Menkalina. Two and a half times as far
to the left, or south, in the direction of Orion, is another
star of equal brightness to Menkalina. This is El Nath, and
marks the place where the foot of Auriga, or the Charioteer,
rests upon the point of the horn of Taurus. Capella, Menkalina, and El Nath make a long triangle which covers the
central part of Auriga. The naked eye shows two or three
misty-looking spots within this triangle, one to the right
of El Nath, one in the upper or eastern part of the constellation, near the third-magnitude star Theta (θ), and another
on a line drawn from Capella to El Nath, but much nearer
to Capella. Turn your glass upon these spots, and you will
be delighted by the beauty of the little stars to whose united
rays they are due.

El Nath has around it some very remarkable rows of
small stars, and the whole constellation of Auriga, like that
of Gemini, glitters with star-dust, for the Milky-Way runs
directly through it.

With a powerful field-glass you may try a glimpse at the rich star-clusters marked 38 M, 37 M, and 33⁷.

The mythology of Auriga is not clear, but the ancients seem to have been of one mind in regarding the constella-

tion as representing the figure of a man carrying a goat and her two kids in his arms. Auriga was also looked upon as a beneficent constellation, and the goat and kids

Map 4.

were believed to be on the watch to rescue shipwrecked sailors. As Capella, which represents the fabled goat, shines nearly overhead in winter, and would ordinarily be the first bright star to beam down through the breaking clouds of a storm at that season, it is not difficult to imagine how it got its reputation as the seaman's friend. Dr. Seiss has so spirited a description of the imaginary figure contained in this constellation that I can not refrain from quoting it :

" The figure itself is that of a mighty man seated on the Milky-Way, holding a band or ribbon in his right hand, and with his left arm holding up on his shoulder a she-goat which clings to his neck and looks out in astonishment upon the terrible bull ; while in his lap are two frightened little kids which he supports with his great hand."

It is scarcely necessary to add that Dr. Seiss insists that Auriga, as a constellation, was invented long before the time of the Greeks, and was intended prophetically to represent that Good Shepherd who was to come and rescue the sinful world.

If any reader wishes to exercise his fancy by trying to trace the outlines of this figure, he will find the head of Auriga marked by the star Delta (δ) and the little group near it. Capella, in the heart of the Goat, is just below his left shoulder, and Menkalina marks his right shoulder. El Nath is in his right foot, and Iota (ι) in his left foot. The stars Epsilon (ε), Zeta (ζ), Eta (η), and Lambda (λ) shine in the kids which lie in Auriga's lap. The faint stars scattered over the eastern part of the constellation are sometimes represented as forming a whip with many lashes, which the giant flourishes with his right hand.

Let us turn back to Denebola in the Lion's Tail. Now glance from it down into the southeast, and you will see a brilliant star flashing well above the horizon. This is Spica. the chief twinkler of Virgo, and it is marked on our circular map. Then look into the northwest, and at about the same distance from Denebola, but higher above the horizon than Spica, you will catch the sparkling of a large, reddish star. It is Arcturus in Boötes. The three, Denebola, Spica, and Arcturus, mark the corners of a great equilateral triangle. Nearly on a line between Denebola and Arcturus, and somewhat nearer to the former, you will perceive a curious twinkling, as if gossamers spangled with dew-drops were entangled there. One might think the old woman of the nursery rhyme who went to sweep the cobwebs out of the sky had skipped this corner, or else that its delicate beauty had preserved it even from her housewifely instincts. This is the little constellation called Berenice's Hair. Your opera-glass will enable you to count twenty or thirty of the largest stars composing this cluster, which are arranged, as so often happens, with a striking appearance of geometrical design. The constellation has a very romantic history. It is related that the young Queen Berenice, when her husband was called away to the wars, vowed to sacrifice her beautiful tresses to Venus if he returned victorious over his enemies.

He did return home in triumph, and Berenice, true to her vow, cut off her hair and bore it to the Temple of Venus. But the same night it disappeared. The king was furious, and the queen wept bitterly over the loss. There is no telling what might have happened to the guardians of the temple, had not a celebrated astronomer named Conon led the young king and queen aside in the evening and showed them the missing locks shining transfigured in the sky. He assured them that Venus had placed Berenice's lustrous ringlets among the stars, and, as they were not skilled in celestial lore, they were quite ready to believe that the silvery swarm they saw near Arcturus had never been there before. And so for centuries the world has recognized the constellation of Berenice's Hair.

Look next at Corvus and Crater, the Crow and the Cup, two little constellations which you will discover on the circular map, and of which we give a separate representation in Map 5. You will find that the stars Delta (δ) and Eta (η), in the upper left-hand corner of the quadrilateral figure of Corvus, make a striking appearance. The little star Zeta (ζ) is a very pretty double for an opera-glass. There is a very faint pair of stars close below and to the right of Beta (β). This forms a severe test. Only a good opera-glass will show these two stars as a single faint point of light. A field-glass, however, will show both, one being considerably fainter than the other. Crater is worth sweeping over for the pretty combinations of stars to be found in it.

You will observe that the interminable Hydra extends his lengthening coils along under both of the constellations. In fact, both the Cup and the Crow are represented as standing upon the huge serpent. The outlines of a cup are tolerably well indicated by the stars included under the name Crater, but the constellation of the Crow might as well have borne any other name so far as any traceable likeness is concerned. One of the legends concerning Corvus avers that it is the daughter of the King of Phocis, who was transformed into a crow to escape the pursuit of Neptune. She is certainly safe in her present guise.

Arcturus and Spica, and their companions, may be left for observation to a more convenient season, when, having risen higher, they can be studied to better advantage. It will be well, however, to merely glance at them with the glass in order to note the great difference of color—Spica being brilliantly white and Arcturus almost red.

MAP 5.

We will now turn to the north. You have already been told how to find the pole-star. Look at it with your glass. The pole-star is a famous double, but its minute companion can only be seen with a telescope. As so often happens, however, it has another companion for the opera-glass, and this latter is sufficiently close and small to make an interesting test for an inexperienced observer armed with a glass of small power. It must be looked for pretty close to the rays of the large star, with such a glass It is of the seventh magnitude. With a large field-glass several smaller companions may be seen, and a very excellent glass may show an 8·5-magnitude star almost hidden in the rays of the seventh-magnitude companion.

With the aid of map No. 6 find in Ursa Minor, which is

the constellation to which the pole-star belongs, the star Beta (β), which is also called Kochab (the star marked *a* in the map is the pole-star). Kochab has a pair of faint stars nearly north of it, about one degree distant. With a small glass these may appear as a single star, but a stronger glass will show them separately.

And now for Ursa Major and the Great Dipper —Draco, Cepheus,

MAP 6.

Cassiopeia, and the other constellations represented on the circular map, being rather too near the horizon for effective observation at this time of the year. First, as the easiest object, look at the star in the middle of the handle of the Dipper (this handle forms the tail of Ursa Major), and a little attention will show you, without the aid of a glass, if your eye-sight is good, that the star is double. A smaller star seems to be almost in contact with it. The larger of these two stars is called Mizar and the smaller Alcor—the Horse and his Rider the Arabs said. Your glass will, of course, greatly increase the distance between Alcor and Mizar, and will also bring out a clear difference of color distinguishing them. Now, if you have a very powerful glass, you may be able to see the Sidus Ludovicianum, a minute star which a German astronomer discovered more than a hundred and fifty years ago, and, strangely enough, taking it for a planet, named it after a German prince. The position

3

of the Sidus Ludovicianum, with reference to Mizar and Alcor, is represented in the accompanying sketch. You must look very sharply if you expect to see it, and your opera-glass will have to be a large and strong one. A field-glass, however, can not fail to show it.

Sweep along the whole length of the Dipper's handle, and you will discover many fine fields of stars. Then look at the star Alpha (*a*) in the outer edge of the bowl nearest to the pole-star. There is a faint star, of about the eighth magnitude, near it, in the direction of Beta (*β*). This will prove a very difficult test. You will have to try it with averted vision. If you have a field-glass, catch it first with that, and, having thus fixed its position in your mind, try to find it with the opera-glass. Its distance is a little over half that between Mizar and Alcor. It is of a reddish color.

You will notice nearly overhead three pairs of pretty bright stars in a long, bending row, about half-way between Leo and the Dipper. These mark three of Ursa Major's feet, and each of the pairs is well worth looking at with a glass, as they are beautifully grouped with stars invisible to the naked eye. The letters used to designate the stars forming these pairs will be found upon our map of Ursa Major. The scattered group of faint stars beyond the bowl of the Dipper forms the Bear's head, and you will find that also a field worth a few minutes' exploration.

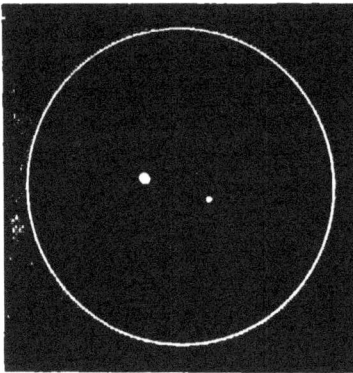

MIZAR, ALCOR, AND THE SIDUS LUDOVICIANUM.

The two bears, Ursa Major and Ursa Minor, swinging around the pole of the heavens, have been conspicuous in the star-lore of all ages. According to fable, they represent the nymph Calisto, with whom Jupiter was in love, and her son Arcas, who were both turned into bears by Juno, whereupon Jupiter, being unable to restore their form, did the next best thing he could by placing them among the stars. Ursa Major is Calisto, or Helica, as the Greeks called the constellation. The Greek name of Ursa Minor was Cynosura. The use of the pole-star in navigation dates back at least to the time of the Phœnicians. The observer will note the uncomfortable position of Ursa Minor, attached to the pole by the end of its long tail.

But, after all, no one can expect to derive from such studies as these any genuine pleasure or satisfaction unless he is mindful of the real meaning of what he sees. The actual truth seems almost too stupendous for belief. The mind must be brought into an attitude of profound contemplation in order to appreciate it. From this globe we can look out in every direction into the open and boundless universe. Blinded and dazzled during the day by the blaze of that star, of which the earth is a near and humble dependent, we are shut in as by a curtain. But at night, when our own star is hidden, our vision ranges into the depths of creation, and we behold them sparkling with a multitude of other suns. With so simple an aid as that of an opera-glass we penetrate still deeper into the profundities of space, and thousands more of these strange, far-away suns come into sight. They are arranged in pairs, sets, rows, streams, clusters—here they gleam alone in distant splendor, there they glow and flash in mighty swarms. This is a look into heaven more splendid than the imagination of Bunyan pictured; here is a celestial city whose temples are suns, and whose streets are the pathways of light.

CHAPTER II.

LET us now suppose that the Earth has advanced for three months in its orbit since we studied the stars of spring, and that, in consequence, the heavens have made one quarter of an apparent revolution. Then we shall find that the stars which in spring shone above the western horizon have been carried down out of sight, while the constellations that were then in the east have now climbed to the zenith, or passed over to the west, and a fresh set of stars has taken their place in the east. In the present chapter we shall deal with what may be called the stars of summer; and, in order to furnish occupation for the observer with an opera-glass throughout the summer months, I have endeavored to so choose the constellations in which our explorations will be made, that some of them shall be favorably situated in each of the months of June, July, and August. The circular map represents the heavens at midnight on the 1st of June ; at eleven o'clock, on the 15th of June ; at ten o'clock, on the 1st of July ; at nine o'clock, on the 15th of July ; and at eight o'clock, on the 1st of August. Remembering that the center of the map is the point over his head, and that the edge of it represents the circle of the horizon, the reader, by a little attention and comparison with the sky, will be able to fix in his mind the relative situation of the various constellations. The maps that follow will show him these constellations on a larger scale, and give him the names of their chief stars.

The observer need not wait until midnight on the 1st of June in order to find some of the constellations included

NORTH.

SOUTH.

MAP 7.

in our map. Earlier in the evening, at about that date, say at nine o'clock, he will be able to see many of these constellations, but he must look for them farther toward the east than they are represented in the map. The bright stars in Boötes and Virgo, for instance, instead of being over in the southwest, as in the map, will be near the meridian ; while Lyra, instead of shining high overhead, will be found climbing up out of the northeast. It would be well to begin at nine o'clock, about the 1st of June, and watch the motions of the heavens for two or three hours. At the commencement of the observations you will find the stars in Boötes, Virgo, and Lyra in the positions I have just mentioned, while half-way down the western sky will be seen the Sickle of Leo. The brilliant Procyon and Capella will be found almost ready to set in the west and northwest, respectively. Between Procyon and Capella, and higher above the horizon, shine the twin stars in Gemini.

In an hour Procyon, Capella, and the Twins will be setting, and Spica will be well past the meridian. In another hour the observer will perceive that the constellations are approaching the places given to them in our map, and at midnight he will find them all in their assigned positions. A single evening spent in observations of this sort will teach him more about the places of the stars than he could learn from a dozen books.

Taking, now, the largest opera-glass you can get (I have before said'that the diameter of the object-glasses should not be less than 1·5 inch, and, I may add, the larger they are the better), find the constellation Scorpio, and its chief star Antares. The map shows you where to look for it at midnight on the 1st of June. If you prefer to begin at nine o'clock at that date, then, instead of looking directly in the south for Scorpio, you must expect to see it just rising in the southeast. You will recognize Antares by its fiery color, as well as by the striking arrangement of its surrounding stars. There are few constellations which bear so close a resemblance to the objects they are named after as Scorpio. It does not require a very violent exercise of the imagination to see in this long, winding trail of stars a gigantic scorpion, with its head to the west, and flourishing its upraised sting that glitters with a pair of twin stars, as if ready to strike. Readers of the old story of Phaeton's disastrous attempt to drive the chariot of the Sun for a day will remember it was the sight of this threatening monster that so terrified the ambitious youth as he dashed along the Zodiac, that he lost control of Apollo's horses, and came near burning the earth up by running the Sun into it.

Antares rather gains in redness when viewed with a glass. Its color is very remarkable, and it is a curious circumstance that with powerful telescopes a small, bright-green star is seen apparently almost touching it. Antares belongs to Secchi's third type of suns, that in which the spectroscopic appearances suggest the existence of a powerfully absorptive atmosphere, and which are believed on various grounds to be, as Lockyer has said, "in the last visible stage of cooling"; in other words, almost extinct. This great, red star probably in actual size exceeds our sun, and no one can help feeling the sublime nature of those studies which give us reason to think that here we can actually behold almost the expiring throes of a giant brother of our giant sun. Only, the lifetime of a sun is many millions of years, and its gradual extinction, even after it has reached a stage as advanced as that of Antares is supposed to be, may occupy a longer time than the whole duration of the human race.

A little close inspection with the naked eye will show three fifth- or sixth-magnitude stars above Antares and Sigma (σ), which form, with those stars, the figure of an irregular pentagon. An opera-glass shows this figure very plainly. The nearest of these stars to Antares, the one directly above it, is known by the number 22, and belongs to Scorpio, while the farthest away, which marks the northernmost corner of the pentagon, is Rho in Ophiuchus. Try a powerful field-glass upon the two stars just named. Take 22 first. You will without much difficulty perceive that it has a little star under its wing, below and to the right, and more than twice as far away above it there is another faint star. Then turn to Rho. Look sharp and you will catch sight of two companion stars, one close to Rho on the right and a little below, and the other still closer and directly above Rho. The latter is quite difficult to be seen distinctly, but the sight is a very pretty one.

The opera-glass will show a number of faint stars scattered around Antares. Turn now to Beta (β) in Scorpio, with the glass. A very pretty pair of stars will be seen hanging below β. Sweeping downward from this point to the horizon you will find many beautiful star-fields. The star marked Nu (ν) is a double which you will be able to separate with a powerful field-glass, the distance between its components being 40″.

And next let us look at a star-cluster. You will see on Map No. 8 an object marked 4 M, near Antares. Its designa-

MAP 8.

tion means that it is No. 4 in Messier's catalogue of nebulæ. It is not a true nebula, but a closely compacted cluster of stars. With the opera-glass, if you are looking in a clear and moonless night, you will see it as a curious nebulous speck. With a field-glass its real nature is more apparent, and it is seen to blaze brighter toward the center. It is, in fact, one of those universes within the universe where thousands of suns are associated together by some unknown law of aggregation into assemblages of whose splendor the slight view that we can get gives us but the faintest conception.

The object above and to the right of Antares, marked in the map 80 M., is a nebula, and although the nebula itself is too small to be seen with an opera-glass (a field-glass shows it as a mere wisp of light), yet there is a pretty array of small stars in its neighborhood worth looking at. Besides, this nebula is of special interest, because in 1860 a star suddenly took its place. At least, that is what seemed to have happened. What really did occur, probably, was that a variable or temporary star, situated between us and the nebula, and ordinarily too faint to be perceived, received a sudden and enormous accession of light, and blazed up so brightly as to blot out of sight the faint nebula behind it. If this star should make its appearance again, it could easily be seen with an opera-glass, and so it will not be useless for the reader to know where to look for it. The quarter of the heavens with which we are now dealing is famous for these celestial conflagrations, if so they may be called. The first temporary star of which there is any record appeared in the constellation of the Scorpion, near the head, 134 years before Christ. It must have been a most extraordinary phenomenon, for it attracted attention all over the world, and both Greek and Chinese annals contain descriptions of it. In 393 A. D. a temporary star shone out in the tail of Scorpio. In 827 A. D. Arabian astronomers, under the Caliph Al-Mamoun, the son of Haroun-al-Raschid, who broke into the great pyramid, observed a temporary star, that shone for four months in the constellation of the Scorpion. In 1203 there was a temporary star, of a bluish color, in the tail of Scorpio, and in 1578 another in the head of the constellation. Besides these there are records of the appearance of four temporary stars in the neighboring constellation of Ophiuchus, one of which, that of 1604, is very famous, and will be described later on. It is conceivable that these strange outbursts in and near Scorpio may have had some effect in causing this constellation to be regarded by the ancients as malign in its influence.

We shall presently see some examples of star-clusters and nebulæ with which the instruments we are using are better capable of dealing than with the one described above. In the mean time, let us follow the bending row of stars from Antares toward the south and east. When you reach the star Mu (μ), you are not unlikely to stop with an exclamation of admiration, for the glass will separate it into two stars that, shining side by side, seem trying to rival each other in brightness. But the next star below μ, marked Zeta (ζ), is even more beautiful. It also separates into two stars, one being reddish and the other bluish in color. The contrast in a clear night is very pleasing. But this is 'not all. Above the two stars you will notice a curious nebulous speck. Now, if you have a powerful field-glass, here is an opportunity to view one of the prettiest sights in the heavens. The field-glass not only makes the two stars appear brighter, and their colors more pronounced, but it shows a third, fainter star below them, making a small triangle, and brings other still fainter stars into sight, while the nebulous speck above turns into a charmingly beautiful little star-cluster, whose components are so close that their rays are inextricably mingled in a maze of light. This little cut is an attempt to represent the scene, but no engraving can reproduce the life and sparkle of it.

ZETA SCORPIONIS.

Following the bend of the Scorpion's tail upward, we come to the pair of stars in the sting. These, of course, are thrown wide apart by the opera-glass. Then let us sweep off to the eastward a little way and find the cluster known as 7 M. You will see it marked on the map. Above it, and near enough to be included in the same field of view, is 6 M., a smaller cluster. Both of these have a sparkling appearance with an opera-glass, and by close attention some of the separate stars in 7 M. may be detected. With a field-glass these clusters become much more striking and starry looking, and the curious radiated structure of 7 M. comes out.

In looking at such objects we can not too often recall to our minds the significance of what we see—that these glimmering specks are the lights in the windows of the universe which carry to us, across inconceivable tracts of space, the assurance that we and our little system are not alone in the heavens; that all around us, and even on the very confines of immensity, Nature is busy, as she is here, and the laws of light, heat, gravitation (and why not of life?), are in full activity.

The clusters we have just been looking at lie on the borders of Scorpio and Sagittarius. Let us cross over into the latter constellation, which commemorates the centaur Chiron. We are now in another, and even a richer, region of wonders. The Milky-Way, streaming down out of the northeast, pours, in a luminous flood, through Sagittarius, inundating that whole region of the heavens with seeming deeps and shallows, and finally bursting the barriers of the horizon disappears, only to glow with redoubled splendor in the southern hemisphere. The stars Zeta (ζ), Tau (τ), Sigma (σ), Phi (ϕ), Lambda (λ), and Mu (μ) indicate the outlines of a figure sometimes called the Milk-Dipper, which is very evident when the eye has once recognized it. On either side of the upturned handle of this dipper-like figure lie some of the most interesting objects in the sky. Let us take the star μ

for a starting-point. Sweep downward and to the right a little
way, and you will be startled by a most singular phenomenon
that has suddenly made its appearance in the field of view of
your glass. You may, perhaps, be tempted to congratulate
yourself on having got ahead of all the astronomers, and discovered a comet. It is really a combination of a star-cluster
with a nebula, and is known as 8 M. Sir John Herschel has
described the "nebulous folds and masses" and dark oval
gaps which he saw in this nebula with his large telescope at
the Cape of Good Hope. But no telescope is needed to make
it appear a wonderful object ; an opera-glass suffices for that,
and a field-glass reveals still more of its marvelous structure.

The reader will recollect that we found the summer solstice close to a wonderful star-swarm in the feet of Gemini.
Singularly enough the winter solstice is also near a star-cluster. It is to be found near a line drawn from 8 M. to the star
μ Sagittarii, and about one third of the way from the cluster
to the star. There is another less conspicuous star-cluster
still closer to the solstitial point here, for this part of the
heavens teems with such aggregations.

On the opposite side of the star μ—that is to say, above
and a little to the left—is an entirely different but almost
equally attractive spectacle, the swarm of stars called 24 M.
Here, again, the field-glass easily shows its superiority over
the opera-glass, for magnifying power is needed to bring out
the innumerable little twinklers of which the cluster is composed. But, whether you use an opera-glass or a field-glass,
do not fail to gaze long and steadily at this island of stars, for
much of its beauty becomes evident only after the eye has
accustomed itself to disentangle the glimmering rays with
which the whole field of view is filled. Try the method of
averted vision, and hundreds of the finest conceivable points
of light will seem to spring into view out of the depths of the
sky. The necessity of a perfectly clear night, and the absence of moonlight, can not be too much insisted upon for

observations such as these. Everybody knows how the moonlight blots out the smaller stars. A slight haziness, or smoke,
in the air produces a similar effect. It is as important to the
observer with an opera-glass to have a transparent atmosphere as it is to one who would use a telescope ; but, fortunately, the work of the former is not so much interfered with
by currents of air. Always avoid the neighborhood of any
bright light. Electric lights in particular are an abomination
to star-gazers.

The cloud of stars we have just been looking at is in a
very rich region of the Milky-Way, in the little modern constellation called "Sobieski's Shield," which we have not
named upon our map. Sweeping slowly upward from 24 M.
a little way with the field-glass, we will pass in succession
over three nebulous-looking spots. The second of these,
counting upward, is the famous Horseshoe nebula. Its wonders are beyond the reach of our instrument, but its place
may be recognized. Look carefully all around this region,
and you will perceive that the old gods, who traveled this
road (the Milky-Way was sometimes called the pathway of
the gods), trod upon golden sands. Off a little way to the
east you will find the rich cluster called 25 M. But do not
imagine the thousands of stars that your opera-glass or field-
glass reveals comprise all the riches of this Golconda of the
heavens. You might ply the powers of the greatest telescope
in a vain attempt to exhaust its wealth. As a hint of the
wonders that lie hidden here, let me quote Father Secchi's
description of a starry spot in this same neighborhood, viewed
with the great telescope at Rome. After telling of "beds of
stars superposed upon one another," and of the wonderful
geometrical arrangement of the larger stars visible in the
field, he adds :

" The greater number are arranged in spiral arcs, in which
one can count as many as ten or twelve stars of the ninth to
the tenth magnitude following one another in a curve, like

beads upon a string. Sometimes they form rays which seem to diverge from a common focus, and, what is very singular, one usually finds, either at the center of the rays, or at the beginning of the curve, a more brilliant star of a red color, which seems to lead the march. It is impossible to believe that such an arrangement can be accidental."

The reader will recall the somewhat similar description that Admiral Smyth and Mr. Webb have given of a star-cluster in Gemini (see Chapter I).

The milky look of the background of the Galaxy is, of course, caused by the intermingled radiations of inconceivably minute and inconceivably numerous stars, thousands of which become separately visible, the number thus distinguishable varying with the size of the instrument. But the most powerful telescope yet placed in human hands can not sound these starry deeps to the bottom. The evidence given by Prof. Holden, the Director of the Lick Observatory, on this point is very interesting. Speaking of the performance of the gigantic telescope on Mount Hamilton, thirty-six inches in aperture, he says :

"The Milky-Way is a wonderful sight, and I have been much interested to see that there is, even with our superlative power, no final resolution of its finer parts into stars. There is always the background of unresolved nebulosity on which hundreds and thousands of stars are studded—each a bright, sharp, separate point."

The groups of stars forming the eastern half of the constellation of Sagittarius are worth sweeping over with the glass, as a number of pretty pairs may be found there.

Sagittarius stands in the old star-maps as a centaur, half-horse-half-man, facing the west, with drawn bow, and arrow pointed at the Scorpion.

Next let us pass to the double constellation adjoining Scorpio and Sagittarius on the north—Ophiuchus and the Serpent. These constellations, as our map shows, are curi-

ously intermixed. The imagination of the old star-gazers, who
named them, saw here the figure of a giant grasping a writh

MAP 9.

ing serpent with his hands. The head of the serpent is
under the Northern Crown, and its tail ends over the star-
gemmed region that we have just described, called "Sobieski's Shield." Ophiuchus stands, as figured in Flamsteed's
"Atlas," upon the back of the Scorpion, holding the serpent
with one hand below the neck, this hand being indicated by
the pair of stars marked Epsilon (ϵ) and Delta (δ), and with
the other near the tail. The stars Tau (τ) and Nu (ν) indicate
the second hand. The giant's face is toward the observer,

and the star Alpha (*a*), also called Ras Alhague, shines in his forehead, while Beta (*β*) and Gamma (*γ*) mark his right shoulder. Ophiuchus has been held to represent the famous physician Æsculapius. One may well repress the tendency to smile at these fanciful legends when he reflects upon their antiquity. There is no doubt that this double constellation is at least three thousand years old—that is to say, for thirty centuries the imagination of men has continued to shape these stars into the figures of a gigantic man struggling with a huge serpent. If it possesses no other interest, then it at least has that which attaches to all things ancient. Like many other of the constellations it has proved longer-lived than the mightiest nations. While Greece flourished and decayed, while Rome rose and fell, while the scepter of civilization has passed from race to race, these starry creations of fancy have shone on unchanged. The mind that would ignore them now deserves compassion.

The reader will observe a little circle in the map, and near it the figures 1604. This indicates the spot where one of the most famous temporary stars on record appeared in the year 1604. At first it was far brighter than any other star in the heavens ; but it quickly faded, and in a little over a year disappeared. It is particularly interesting, because Kepler—the quaintest, and not far from the greatest, figure in astronomical history—wrote a curious book about it. Some of the philosophers of the day argued that the sudden outburst of the wonderful star was caused by the chance meeting of atoms. Kepler's reply was characteristic, as well as amusing :

"I will tell those disputants, my opponents, not my own opinion, but my wife's. Yesterday, when I was weary with writing, my mind being quite dusty with considering these atoms, I was called to supper, and a salad I had asked for was set before me. 'It seems, then,' said I, aloud, 'that if pewter dishes, leaves of lettuce, grains of salt, drops of water, vinegar and oil, and slices of egg, had been flying about in the air from all eternity, it might at last happen by chance that there would come a salad.' 'Yes,' says my wife, 'but not so nice and well-dressed as this of mine is.' "

While there are no objects of special interest for the observer with an opera-glass in Ophiuchus, he will find it worth while to sweep over it for what he may pick up, and, in particular, he should look at the group of stars southeast of β and γ. These stars have been shaped into a little modern asterism called Taurus Poniatowskii, and it will be noticed that five of them mark the outlines of a letter V, resembling the well-known figure of the Hyades.

Also look at the stars in the head of Serpens, several of which form a figure like a letter X. A little west of Theta (θ), in the tail of Serpens, is a beautiful swarm of little stars, upon which a field-glass may be used with advantage. The star θ is itself a charming double, just within the separating power of a very powerful field-glass under favorable circumstances, the component stars being only about one third of a minute apart.

Do not fail to notice the remarkable subdivisions of the Milky-Way in this neighborhood. Its current seems divided into numerous channels and bays, interspersed with gaps that might be likened to islands, and the star θ appears to be situated upon one of these islands of the galaxy. This complicated structure of the Milky-Way extends downward to the horizon, and upward through the constellation Cygnus, and of its phenomenal appearance in that region we shall have more to say further on.

Directly north of Ophiuchus is the constellation Hercules, interesting as occupying that part of the heavens toward which the proper motion of the sun is bearing the earth and its fellow-planets, at the rate, probably, of not less than 160,-000,000 miles in a year—a stupendous voyage through space, of whose destination we are as ignorant as the crew of a ship sailing under sealed orders, and, like whom, we must depend

4

upon such inferences as we can draw from courses and distances, for no other information comes to us from the flag-
ship of our squadron.

In the accompanying map we have represented the beautiful constellations Lyra and the Northern Crown, lying on

MAP 10.

either side of Hercules. The reader should note that the
point overhead in this map is not far from the star Eta (η) in
Hercules. The bottom of the map is toward the south, the
right-hand side is west, and the left-hand side east. It is important to keep these directions in mind, in comparing the
map with the sky. For instance, the observer must not expect to look into the south and see Hercules half-way up the
sky, with Lyra a little east of it ; he must look for Hercules

nearly overhead, and Lyra a little east of the zenith. The same precautions are not necessary in using the maps of Scorpio, Sagittarius, and Ophiuchus, because those constellations are nearer the horizon, and so the observer does not have to imagine the map as being suspended over his head.

The name Hercules sufficiently indicates the mythological origin of the constellation, and yet the Greeks did not know it by that name, for Aratus calls it "the Phantom whose name none can tell." The Northern Crown, according to fable, was the celebrated crown of Ariadne, and Lyra was the harp of Orpheus himself, with whose sweet music he charmed the hosts of Hades, and persuaded Pluto to yield up to him his lost Eurydice.

With the aid of the map you will be able to recognize the principal stars and star-groups in Hercules, and will find many interesting combinations of stars for yourself. An object of special interest is the celebrated star-cluster 13 M. You will find it on the map between the stars Eta (η) and Zeta (ζ). While an opera-glass will only show it as a faint and minute speck, lying nearly between two little stars, it is nevertheless well worth looking for, on account of the great renown of this wonderful congregation of stars. Sir William Herschel computed the number of stars contained in it as about fourteen thousand. It is roughly spherical in shape, though there are many straggling stars around it evidently connected with the cluster. In short, it is *a ball of suns*. The reader should not mistake what that implies, however. These suns, though truly solar bodies, are probably very much smaller than our sun. Mr. Gore has computed their average diameter to be forty-five thousand miles, and the distance separating each from the next to be 9,000,000,000 miles. It may not be uninteresting to inquire what would be the appearance of the sky to dwellers within such a system of suns. Adopting Mr. Gore's estimates, and supposing 9,000,000,000 miles to be very nearly the uniform distance apart of the stars in the cluster, and forty-five thousand miles their uniform diameter, then, starting with a single star in the center, their arrangement might be approximately in concentric spherical shells, situated about 9,000,000,000 miles apart. The first shell, counting outward from the center, would contain a dozen stars, each of which, as seen by an observer stationed upon a planet at the center of the cluster, would shine eleven hundred times as bright as Sirius appears to us. The number of the stars in each shell would increase as they receded from the center in proportion to the squares of the radii of the successive shells, while their luminosity, as seen from the center, would vary inversely as those squares. Still, the outermost stars—the total number being limited to fourteen or fifteen thousand—would appear to our observer at the center of the system about five times as brilliant as Sirius.

It is clear, then, that he would be dwelling in a sort of perpetual daylight. His planet might receive from the particular sun around which it revolved as brilliant a daylight as our sun gives to us, but let us see what would be the illumination of its night side. Adopting Zöllner's estimate of the light of the sun as 618,000 times as great as that of the full moon, and choosing among the various estimates of the light of Sirius as compared with the sun $\frac{1}{40000000000}$ as probably the nearest the truth, we find that the moon sends us about sixty-five hundred times as much light as Sirius does. Now, since the dozen stars nearest the center of the cluster would each appear to our observer eleven hundred times as bright as Sirius, all of them together would give a little more than twice as much light as the full moon sheds upon the earth. But as only half the stars in the cluster would be above the horizon at once we must diminish this estimate by one half, in order to obtain the amount of light that our supposititious planet would receive on its night side from the nearest stars in the cluster. And since the number of these stars increases

with their distance from the center in the same ratio as their
light diminishes, it follows that the total light received from
the cluster would exceed that received from the dozen nearest
stars as many times as there were spherical shells in the cluster. This would be about fifteen times, and accordingly all
the stars together would shed, at the center, some thirty times
as much light as that of the moon. Dividing this again by
two, because only half of the stars could be seen at once, we
find that the night side of our observer's planet would be illuminated with fifteen times as much light as the full moon
sheds upon the earth.

It is evident, too, that our observer would enjoy the spectacle of a starry firmament incomparably more splendid than
that which we behold. Only about three thousand stars are
visible to our unassisted eyes at once on any clear night, and
of those only a few are conspicuous, and two thirds are so
faint that they require some attention in order to be distinguished. But the spectator at the center of the Hercules
cluster would behold some seven thousand stars at once, the
faintest of which would be five times as brilliant as the brightest star in our sky, while the brighter ones would blaze like
nearing suns. One effect of this flood of starlight would be
to shut out from our observer's eyes all the stars of the outside universe. They would be effaced in the blaze of his sky,
and he would be, in a manner, shut up within his own little
star-system, knowing nothing of the greater universe beyond,
in which we behold his multitude of luminaries, diminished
and blended by distance into a faintly shining speck, floating
like a silvery mote in a sunbeam.

If our observer's planet, instead of being situated in the
center of the cluster, circled around one of the stars at the
outer edge of it, the appearance of his sky would be, in some
respects, still more wonderful, the precise phenomena de-
pending upon the position of the planet's orbit and the sta-
tion of the observer. Less than half of his sky would be

filled, at any time, by the stars of the cluster, the other half opening upon outer space and appearing by comparison almost starless—a vast, cavernous expanse, with a few faint glimmerings out of its gloomy depths. The plane of the orbit of his planet being supposed to pass through the center of the spherical system, our observer would, during his year, behold the night at one season blazing with the splendors of the clustered suns, and at another emptied of brilliant orbs and faintly lighted with the soft glow of the Milky-Way and the feeble flickering of distant stars, scattered over the dark vault. The position of the orbit, and the inclination of the planet's axis might be such that the glories of the cluster would not be visible from one of its hemispheres, necessitating a journey to the other side of the globe to behold them.*

Of course, it is not to be assumed that the arrangement of the stars in the cluster actually is exactly that which we have imagined. Still, whatever the arrangement, so long as the cluster is practically spherical, and the stars composing it are of nearly uniform size and situated at nearly uniform distances, the phenomena we have described would fairly represent the appearances presented to inhabitants of worlds situated in such a system. As to the possibility of the existence of such worlds and inhabitants, everybody must draw his own conclusions. Astronomy, as a science, is silent upon that question. But there shine the congregated stars, mingling their rays in a message of light, that comes to us across the gulf, proclaiming their brotherhood with our own glorious sun. Mathematicians can not unravel the interlocking intricacies of their orbits, and some would, perhaps *a priori*, have said that such a system was impossible, but the telescope has revealed them, and there they are! What purposes they subserve in the economy of the universe, who shall declare?

* A similar calculation of the internal appearances of the Hercules cluster, which I made, was published in 1887 in the " New York Sun."

If you have a field-glass, by all means try it upon 13 M. It will give you a more satisfactory view than an opera-glass is capable of doing, and will magnify the cluster so that there can be no possibility of mistaking it for a star. Compare this compact cluster, which only a powerful telescope can partially resolve into its component stars, with 7 M. and 24 M., described before, in order to comprehend the wide variety in the structure of these aggregations of stars.

The Northern Crown, although a strikingly beautiful constellation to the naked eye, offers few attractions to the opera-glass. Let us turn, then, to Lyra. I have never been able to make up my mind which of three great stars is entitled to precedence—Vega, the leading brilliant of Lyra, Arcturus in Boötes, or Capella in Auriga. They are the three leaders of the northern firmament, but which of them should be called the chief, is very hard to say. At any rate, Vega would probably be generally regarded as the most beautiful, on account of the delicate bluish tinge in its light, especially when viewed with a glass. There is no possibility of mistaking this star because of its surpassing brilliancy. Two faint stars close to Vega on the east make a beautiful little triangle with it, and thus form a further means of recognition, if any were needed. Your opera-glass will show that the floor of heaven is powdered with stars, fine as the dust of a diamond, all around the neighborhood of Vega, and the longer you gaze the more of these diminutive twinklers you will discover.

Now direct your glass to the northernmost of the two little stars near Vega, the one marked Epsilon (ϵ) in the map. You will perceive that it is composed of two stars of almost equal magnitude. If you had a telescope of considerable power, you would find that each of these stars is in turn double. In other words, this wonderful star which appears single to the unassisted eye, is in reality quadruple, and there is reason to think that the four stars composing it are connected in pairs, the members of each pair revolving around their common center while the two pairs in turn cir-

MAP 11.

cle around a center common to all. With a field-glass you will be able to see that the other star near Vega, Zeta (ζ), is also double, the distance between its components being three quarters of a minute, while the two stars in ϵ are a little less than $3\frac{1}{2}'$ apart. The star Beta (β) is remarkably variable in brightness. You may watch these variations, which run through a regular period of about 12 days, $21\frac{3}{4}$ hours, for yourself. Between Beta and Gamma (γ) lies the beautiful Ring nebula, but it is hopelessly beyond the reach of the optical means we are employing.

Let us turn next to the stars in the west. In consulting the accompanying map of Virgo and Boötes (Map No. 11), the observer is supposed to face the southwest, at the hours and dates mentioned above as those to which the circular map corresponds. He will then see the bright star Spica in Virgo not far above the horizon, while Arcturus will be half-way up the sky, and the Northern Crown will be near the zenith.

The constellation Virgo is an interesting one in mythological story. Aratus tells us that the Virgin's home was once on earth, where she bore the name of Justice, and in the golden age all men obeyed her. In the silver age her visits to men became less frequent, "no longer finding the spirits of former days"; and, finally, when the brazen age came with the clangor of war:

> " Justice, loathing that race of men,
> Winged her flight to heaven ; and fixed
> Her station in that region
> Where still by night is seen
> The Virgin goddess near to bright Boötes."

The chief star of Virgo, Spica, is remarkable for its pure white light. To my eye there is no conspicuous star in the sky equal to it in this respect, and it gains in beauty when viewed with a glass. With the aid of the map the reader will find the celebrated binary star Gamma (γ) Virginis, although he will not be able to separate its components without a telescope. It is a curious fact that the star Epsilon (ϵ) in Virgo has for many ages been known as the Grape-Gatherer. It has borne this name in Greek, in Latin, in Persian, and in Arabic, the origin of the appellation undoubtedly being that it was observed to rise just before the sun in the season of the vintage. It will be observed that the stars ϵ, δ, γ, η, and β, mark two sides of a quadrilateral figure of which the opposite corner is indicated by Denebola in the tail of Leo. Within this quadrilateral lies the marvelous Field of the

Nebulæ, a region where with adequate optical power one may find hundreds of these strange objects thronging together, a very storehouse of the germs of suns and worlds. Unfortunately, these nebulæ are far beyond the reach of an opera-glass, but it is worth while to know where this curious region is, even if we can not behold the wonders it contains. The stars Omicron (o), Pi (π), etc., forming a little group, mark the head of Virgo.

The autumnal equinox, or the place where the sun crosses the equator of the heavens on his southerly journey about the 21st of September, is situated nearly between the stars η and β Virginis, a little below the line joining them, and somewhat nearer to η. Both η and ζ Virginis are almost exactly upon the equator of the heavens.

The constellation Libra, lying between Virgo and Scorpio, does not contain much to attract our attention. Its two chief stars, a and β, may be readily recognized west of and above the head of Scorpio. The upper one of the two, β, has a singular greenish tint, and the lower one, a, is a very pretty double for an opera-glass.

The constellation of Libra appears to have been of later date than the other eleven members of the zodiacal circle. Its two chief stars at one time marked the extended claws of Scorpio, which were afterward cut off (perhaps the monster proved too horrible even for its inventors) to form Libra. As its name signifies, Libra represents a balance, and this fact seems to refer the invention of the constellation back to at least three hundred years before Christ, when the autumnal equinox occurred at the moment when the sun was just crossing the western border of the constellation. The equality of the days and nights at that season readily suggests the idea of a balance. Milton, in "Paradise Lost," suggests another origin for the constellation of the Balance in the account of Gabriel's discovery of Satan in paradise :

" . . . Now dreadful deeds
Might have ensued, nor only paradise
In this commotion, but the starry cope
Of heaven, perhaps, or all the elements
At least had gone to wrack, disturbed and torn
With violence of this conflict, had not soon
The Eternal, to prevent such horrid fray,
Hung forth in heaven his golden scales, yet seen
Betwixt Astrea and the Scorpion sign."

Just north of Virgo's head will be seen the glimmering of Berenice's Hair. This little constellation was included among those described in the chapter on "The Stars of Spring," but it is worth looking at again in the early summer, on moonless nights, when the singular arrangement of the brighter members of the cluster at once strikes the eye.

Boötes, whose leading brilliant, Arcturus, occupies the center of our map, also possesses a curious mythical history. It is called by the Greeks the Bear-Driver,

BERENICE'S HAIR.

because it seems continually to chase Ursa Major, the Great Bear, in his path around the pole. The story is that Boötes was the son of the nymph Calisto, whom Juno, in one of her customary fits of jealousy, turned into a bear. Boötes, who had become a famous hunter, one day roused a bear from her lair, and, not knowing that it was his mother, was about to kill her, when Jupiter came to the rescue and snatched them both up into the sky, where they have shone ever since. Lucan refers to this story when, describing Brutus's visit to Cato at night, he fixes the time by the position of these constellations in the heavens :

" 'Twas when the solemn dead of night came on,
When bright Calisto, with her shining son,
Now half the circle round the pole had run."

Boötes is not specially interesting for our purposes, except for the splendor of Arcturus. This star has possessed a peculiar charm for me ever since boyhood, when, having read a description of it in an old treatise on Uranography, I felt an eager desire to see it. As my search for it chanced to begin at a season when Arcturus did not rise till after a boy's bed-time, I was for a long time disappointed, and I shall never forget the start of surprise and almost of awe with which I finally caught sight of it, one spring evening, shooting its flaming rays through the boughs of an apple-orchard, like a star on fire.

When near the horizon, Arcturus has a remarkably reddish color; but, after it has attained a high elevation in the sky, it appears rather a deep yellow than red. There is a scattered cluster of small stars surrounding Arcturus, forming an admirable spectacle with an opera-glass on a clear night. To see these stars well, the glass should be slowly moved about. Many of them are hidden by the glare of Arcturus. The little group of stars near the end of the handle of the Great Dipper, or, what is the same thing, the tail of the Great Bear, marks the upraised hand of Boötes. Between Berenice's Hair and the tail of the Bear you will see a small constellation called Canes Venatici, the Hunting-Dogs. On the old star-maps Boötes is represented as holding these dogs with a leash, while they are straining in chase of the Bear. You will find some pretty groupings of stars in this constellation.

And now we will turn to the east. Our next map shows Cygnus, a constellation especially remarkable for the large and striking figure that it contains, called the Northern Cross, Aquila the Eagle, the Dolphin, and the little asterisms Sagitta and Vulpecula. In consulting the map, the observer is supposed to face toward the east. In Aquila the curious arrangement of two stars on either side of the chief star of the constellation, called Altair, at once attracts the eye. Within a circle including the two attendants of Altair you will probably be able to see with the naked eye only two or three stars in addition to the three large ones. Now turn your glass upon the same spot, and you will see eight or ten times as many stars, and with a field-glass still more can be seen. Watch the star marked Eta (η), and you will find that its light is variable, being sometimes more than twice as bright as at other times. Its changes are periodical, and occupy a little over a week.

The Eagle is fabled to have been the bird that Jupiter kept beside his throne. A constellation called Antinous, invented by Tycho Brahe, is represented on some maps as occupying the lower portion of the space given to Aquila.

The Dolphin is an interesting little constellation, and the ancients said it represented the very animal on whose back the famous musician Arion rode through the sea after his escape from the sailors who tried to murder him. But some modern has dubbed it with the less romantic name of Job's Coffin, by which it is sometimes called. It presents a very pretty sight to the opera-glass.

Cygnus, the swan, is a constellation whose mythological history is not specially interesting, although, as remarked above, it contains one of the most clearly marked figures to be found among the stars, the famous Northern Cross. The outlines of this cross are marked with great distinctness by the stars Alpha (a), Epsilon (ϵ), Gamma (γ), Delta (δ), and Beta (β), together with some fainter stars lying along the main beam of the cross between β and γ. The star β, also called Albireo, is one of the most beautiful double stars in the heavens. The components are sharply contrasted in color, the larger star being golden-yellow, while the smaller one is a deep, rich blue. With a field-glass of 1·6-inch aperture and magnifying seven times I have sometimes been able to divide this pair, and to recognize the blue color of the smaller star. It will be found a severe test for such a glass.

MAP 12.

About half-way from Albireo to the two stars ζ and ε in Aquila is a very curious little group, consisting of six or seven stars in a straight row, with a garland of other stars hanging from the center. To see it best, take a field-glass, although an opera-glass shows it.

I have indicated the place of the celebrated star 61 Cygni in the map, because of the interest attaching to it as the nearest to us, so far as we know, of all the stars in the northern hemisphere, and with one exception the nearest star in all the heavens. Yet it is very faint, and the fact that so inconspicuous a star should be nearer than such brilliants as Vega and Arcturus shows how wide is the range of magnitude among the suns that light the universe. The actual distance of 61 Cygni is something like 650,000 times as great as the distance from the earth to the sun.

The star Omicron (*o*) is very interesting with an operaglass. The naked eye sees a little star near it. The glass throws them wide apart, and divides *o* itself into two stars. Now, a field-glass, if of sufficient power, will divide the larger of these stars again into two—a fine test.

Sweep around *a* and *γ* for the splendid star-fields that abound in this neighborhood ; also around the upper part of the figure of the cross. We are here in one of the richest parts of the Milky-Way. Between the stars *a*, *γ*, *ε*, is the strange dark gap in the galaxy called the Coal-Sack, a sort of hole in the starry heavens. Although it is not entirely empty of stars, its blackness is striking in contrast with the brilliancy of the Milky-Way in this neighborhood. The divergent streams of the great river of light in this region present a very remarkable appearance.

Finally, we come to the great dragon of the sky. In using the map of Draco and the neighboring constellations, the reader is supposed to face the north. The center of the upper edge of the map is directly over the observer's head. One of the stories told of this large constellation is that it represents a dragon that had the temerity to war against Minerva. The goddess "seized it in her hand, and hurled it, twisted as it was, into the heavens round the axis of the world, before it had time to unwind its contortions." Others say it is the dragon that guarded the golden apples in the Garden of the Hesperides, and that was slain by the redoubtable Hercules. At any rate, it is plainly a monster of the first magnitude. The stars *β*, *γ*, *ξ*, *ν*, and *μ* represent its head, while its

body runs trailing along, first sweeping in a long curve toward Cepheus, and then bending around and passing between

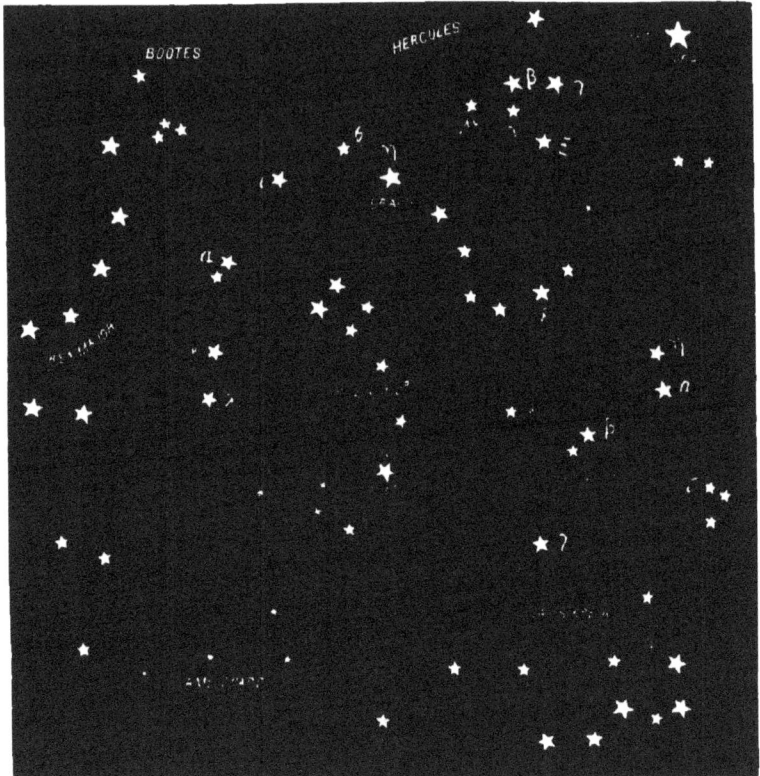

MAP 13.

the two bears.   Try ν with your opera-glass, and if you succeed in seeing it double you may congratulate yourself on
your keen sight.   The distance between the stars is about 1′.
Notice the contrasted colors of γ and β, the former being a
rich orange and the latter white.   As you sweep along the
winding way that Draco follows, you will run across many
striking fields of stars, although the heavens are not as rich
here as in the splendid regions that we have just left.   You
will also find that Cepheus, although not an attractive con-

stellation to the naked eye, is worth some attention with an opera-glass. The head and upper part of the body of Cepheus are plunged in the stream of the Milky Way, while his feet are directed toward the pole of the heavens, upon which he is pictured as standing. Cepheus, however, sinks into insignificance in comparison with its neighbor Cassiopeia, but that constellation belongs rather to the autumn sky, and we shall pass it by here.

CHAPTER III.

IN the "Fifth Evening" of that delightful, old, out-of-date book of Fontenelle's, on the "Plurality of Worlds," the Astronomer and the Marchioness, who have been making a wonderful pilgrimage through the heavens during their evening strolls in the park, come at last to the starry systems beyond the "solar vortex," and the Marchioness experiences a lively impatience to know what the fixed stars will turn out to be, for the Astronomer has sharpened her appetite for marvels.

"Tell me," says she, eagerly, "are they, too, inhabited like the planets, or are they not peopled? In short, what can we make of them?"

The Astronomer answers his charming questioner, as we should do to-day, that the fixed stars are so many suns. And he adds to this information a great deal of entertaining talk about the planets that may be supposed to circle around these distant suns, interspersing his conversation with explanations of "vortexes," and many quaint conceits, in which he is helped out by the ready wit of the Marchioness.

Finally, the impressionable mind of the lady is overwhelmed by the grandeur of the scenes that the Astronomer opens to her view, her head swims, infinity oppresses her, and she cries for mercy.

"You show me," she exclaims, "a perspective so interminably long that the eye can not see the end of it. I see plainly the inhabitants of the earth; then you cause me to

perceive those of the moon and of the other planets belonging
to our vortex (system), quite clearly, yet not so distinctly as
those of the earth. After them come the inhabitants of planets in the other vortexes. I confess, they seem to me hidden
deep in the background, and, however hard I try, I can bare-
ly glimpse them at all. In truth, are they not almost anni-
hilated by the very expression which you are obliged to use
in speaking of them? You have to call them inhabitants of
one of the planets contained in one out of the infinity of vortexes. Surely we ourselves, to whom the same expression
applies, are almost lost among so many millions of worlds.
For my part, the earth begins to appear so frightfully little
to me that henceforth I shall hardly consider any object worthy of eager pursuit. Assuredly, people who seek so earnest-
ly their own aggrandizement, who lay schemes upon schemes,
and give themselves so much trouble, know nothing of the
vortexes! I am sure my increase of knowledge will redound
to the credit of my idleness, and when people reproach me
with indolence I shall reply : 'Ah! if you but knew the history of the fixed stars ! ' "

It is certainly true that a contemplation of the unthink-
able vastness of the universe, in the midst of which we dwell
upon a speck illuminated by a spark, is calculated to make all
terrestrial affairs appear contemptibly insignificant. We can
not wonder that men for ages regarded the earth as the center, and the heavens with their lights as tributary to it, for to
have thought otherwise, in those times, would have been to
see things from the point of view of a superior intelligence.
It has taken a vast amount of experience and knowledge to
convince men of the parvitude of themselves and their belong-
ings. So, in all ages they have applied a terrestrial measure
to the universe, and imagined they could behold human
affairs reflected in the heavens and human interests setting the
gods together by the ears.

This is clearly shown in the story of the constellations.

The tremendous truth that on a starry night we look, in every direction, into an almost endless vista of suns beyond

MAP. 14.

suns and systems upon systems, was too overwhelming for comprehension by the inventors of the constellations. So they amused themselves, like imaginative children, as they were, by tracing the outlines of men and beasts formed by those pretty lights, the stars. They turned the starry heavens into a scroll filled with pictured stories of mythology.

Four of the constellations with which we are going to deal in this chapter are particularly interesting on this account. They preserve in the stars, more lasting than parchment or stone, one of the oldest and most pleasing of all the romantic stories that have amused and inspired the minds of men—the story of Perseus and Andromeda—a better story than any that modern novelists have invented. The four constellations to which I refer bear the names of Andromeda, Perseus, Cassiopeia, and Cepheus, and are sometimes called, collectively, the Royal Family. In the autumn they occupy a conspicuous position in the sky, forming a group that remains unrivaled until the rising of Orion with his imperial *cortége.* The reader will find them in Map No. 14, occupying the northeastern quarter of the heavens.

This map represents the visible heavens at about midnight on September 1st, ten o'clock P. M. on October 1st, and eight o'clock P. M. on November 1st. At this time the constellations that were near the meridian in summer will be found sinking in the west, Hercules being low in the northwest, with the brilliant Lyra and the head of Draco suspended above it; Aquila, "the eagle of the winds," soars high in the southwest; while the Cross of Cygnus is just west of the zenith; and Sagittarius, with its wealth of star-dust, is disappearing under the horizon in the southwest.

Far down in the south the observer catches the gleam of a bright lone star of the first magnitude, though not one of the largest of that class. It is Fomalhaut, in the mouth of the Southern Fish, Piscis Australis. A slight reddish tint will be perceived in the light of this beautiful star, whose brilliance is enhanced by the fact that it shines without a rival in that region of the sky. Fomalhaut is one of the important "nautical stars," and its position was long ago carefully computed for the benefit of mariners. The constellation of Piscis Australis, which will be found in our second map, does not possess much to interest us except its splendid leading star.

In consulting Map 15, the observer is supposed to be facing south, or slightly west of south, and he must remember that the upper part of the map reaches nearly to the zenith, while at the bottom it extends down to the horizon.

To the right, or west, of Fomalhaut, and higher up, is the constellation of Capricornus, very interesting on many accounts, though by no means a striking constellation to the unassisted eye. The stars Alpha (a), called Giedi, and Beta (β), called Dabih, will be readily recognized, and a keen eye will perceive that Alpha really consists of two stars. They are about six minutes of arc apart, and are of the third and the fourth magnitude respectively. These stars, which to the naked eye appear almost blended into one, really have no physical connection with each other, and are slowly drifting apart. The ancient astronomers make no mention of Giedi being composed of two stars, and the reason is plain, when it is known that in the time of Hipparchus, as Flammarion has pointed out, their distance apart was not more than two thirds as great as it is at present, so that the naked eye could not have detected the fact that there were two of them ; and it was not until the seventeenth century that they got far enough asunder to begin to be separated by eyes of unusual power. With an ordinary opera-glass they are thrown well apart, and present a very pretty sight. Considering the manner in which these stars are separating, the fact that both of them have several faint companions, which our powerful telescopes reveal, becomes all the more interesting. A suggestion of Sir John Herschel, concerning one of these faint companions, that it shines by reflected light, adds to the interest, for if the suggestion is well founded the little star must, of course, be actually a planet, and granting that, then some of the other faint points of light seen there are probably planets too. It must be said that the probabilities are against Herschel's suggestion. The faint stars more likely shine with their own light. Even so, however, these two systems, which apparently have met and are passing one another, at a distance small as compared with the space that separates them from us, possess a peculiar interest, like two celestial fleets that have spoken one another in the midst of the ocean of space.

The star Beta, or Dabih, is also a double star. The companion is of a beautiful blue color, generally described as "sky-blue." It is of the seventh magnitude, while the larger star is of magnitude three and a half. The latter is golden-yellow. The blue of the small star can be seen with either an opera- or a field-glass, but it requires careful looking and a clear and steady atmosphere. I recollect discovering the color of this star with a field-glass, and exclaiming to myself, "Why, the little one is as blue as a bluebell!" before I knew that that was its hue as seen with a telescope. Trying my opera-glass upon it I found that the color was even more distinct, although the small star was then more or less enveloped in the yellow rays of the large one. The distance between the two stars in Dabih is nearly the same as that between the components of ε Lyræ, and the comparative difficulty of separating them is an instructive example of the effect of a large star in concealing a small one close beside it. The two stars in ε Lyræ are of nearly equal brightness, and are very easily separated and distinguished, but in β Capricorni, or Dabih, one star is about twenty times as bright as the other, and consequently the fainter star is almost concealed in the glare of its more brilliant neighbor.

With the most powerful glass at your disposal, sweep from the star Zeta (ζ) eastward a distance somewhat greater than that separating Alpha and Beta, and you will find a fifth-magnitude star beside a little nebulous spot. This is the cluster known as 30 M, one of those sun-swarms that overwhelm the mind of the contemplative observer with astonishment, and especially remarkable in this case for the apparent vacancy of the heavens immediately surrounding the cluster, as if all the stars in that neighborhood had been drawn into the great assemblage, leaving a void around it. Of course, with the instrument that our observer is supposed to be using, merely the *existence* of this solar throng can be detected ; but, if he sees that it is there, he may be led to provide himself with a telescope capable of revealing its glories.

Admiral Smyth remarks that, "although Capricorn is not a striking object, it has been the very pet of all constellations with astrologers," and he quotes from an old almanac of the year 1386, that "whoso is borne in Capcorn schal be ryche and wel lufyd." The mythological account of the constellation is that it represents the goat into which Pan was turned in order to escape from the giant Typhon, who once on a time scared all the gods out of their wits, and caused them to change themselves into animals, even Jupiter assuming the form of a ram. According to some authorities, Piscis Australis represents the fish into which Venus changed herself on that interesting occasion.

Directly above Piscis Australis, and to the east or left of Capricorn, the map shows the constellation of Aquarius, or the Water-Bearer. Some say this commemorates Ganymede, the cup-bearer of the gods. It is represented in old star-maps by the figure of a young man pouring water from an urn. The star Alpha (*a*) marks his right shoulder, and Beta (*β*) his left, and Gamma (*γ*), Zeta (*ζ*), Eta (*η*), and Pi (*π*) indicate his right hand and the urn. From this group a current of small stars will be recognized, sweeping downward with a curve toward the east, and ending at Fomalhaut; this represents the water poured from the urn, which the Southern Fish appears to be drinking. In fact, according to the pictures in the old maps, the fish succeeds in swallowing the stream completely, and it vanishes from the sky in the act of entering his distended mouth! It is worthy of remark that in Greek, Latin, and Arabic this constellation bears names all of which signify "a man pouring water." The ancient Egyptians imagined that the setting of Aquarius caused the rising of the Nile, as he sank his huge urn in the river to fill it. Alpha Aquarii was called by the Arabs Sadalmelik, which is interpreted to mean the "king's lucky star," but whether it proved itself a lucky star in war or in love, and what particular king enjoyed its benign influence and recorded his gratitude in its name, we are not informed. Thus, at every step, we find how shreds of history and bits of superstition are entangled among the stars. Surely, humanity has been reflected in the heavens as lastingly as it has impressed itself upon the earth.

Starting from the group of stars just described as forming the Water-Bearer's urn, follow with a glass the winding stream of small stars that represent the water. Several very pretty and striking assemblages of stars will be encountered in its course. The star Tau (τ) is double and presents a beautiful contrast of color, one star being white and the other reddish-orange—two solar systems, it may be, apparently neighbors as seen from the earth, in one of which daylight is white and in the other red !

Point a good glass upon the star marked Nu (ν), and you will see, somewhat less than a degree and a half to the west of it, what appears to be a faint star of between the seventh and eighth magnitudes. You will have to look sharp to see it. It is with your mind's eye that you must gaze, in order to perceive the wonder here hidden in the depths of space. That faint speck is a nebula, unrivaled for interest by many of the larger and more conspicuous objects of that kind. Lord Rosse's great telescope has shown that in form it resembles the planet Saturn ; in other words, that it consists apparently of a ball surrounded by a ring. But the spectroscope proves that it is a gaseous mass, and the micrometer— supposing its distance to be equal to that of the stars, and we have no reason to think it less—that it must be large enough to fill the whole space included within the orbit of Neptune ! Here, then, as has been said, we seem to behold a genesis in the heavens. If Laplace's nebular hypothesis, or any of the modifications of that hypothesis, represents the process of formation of a solar system, then we may fairly conclude that such a process is now actually in operation in this nebula in Aquarius, where a vast ring of nebulous matter appears to have separated off from the spherical mass within it. This may not be the true explanation of what we see there, but, whatever the explanation is, there can be no question of the high significance of this nebula, whose shape proclaims unmistakably the operation of great metamorphic forces there. Of course, with his insignificant optical means, our observer can see nothing of the strange form of this object, the detection of which requires the aid of the most powerful telescopes, but it is much to know where that unfinished creation lies, and to see it, even though diminished by distance to a mere speck of light.

Turn your glass upon the star shown in the map just above Mu (μ) and Epsilon (ε). You will find an attractive arrangement of small stars in its neighborhood. The star marked 104 is double to the naked eye, and the row of stars below it is well worth looking at. The star Delta (δ) indicates the place where, in 1756, Tobias Mayer narrowly escaped making a discovery that would have anticipated that which a quarter of a century later made the name of Sir William Herschel world-renowned. The planet Uranus passed near Delta in 1756, and Tobias Mayer saw it, but it moved so slowly that he took it for a fixed star, never suspecting that his eyes had rested upon a member of the solar system whose existence was, up to that time, unknown to the inhabitants of Adam's planet.

Above Aquarius you will find the constellation Pegasus. It is conspicuously marked by four stars of about the second magnitude, which shine at the corners of a large square, called the Great Square of Pegasus. This figure is some fifteen degrees square, and at once attracts the eye, there being few stars visible within the quadrilateral, and no large ones in the immediate neighborhood to distract attention from it. One of the four stars, however, as will be seen by consulting Map 15, does not belong to Pegasus, but to the constellation Andromeda. Mythologically, this constellation represents the celebrated winged horse of antiquity :

" Now heaven his further wandering flight confines,
Where, splendid with his numerous stars, he shines."

The star Alpha (α) is called Markab; Beta (β) is Scheat,
and Gamma (γ) is Algenib; the fourth star in the square,
belonging to Andromeda, is called Alpheratz. Although
Pegasus presents a striking appearance to the unassisted eye,
on account of its great square, it contains little to attract the
observer with an opera-glass. It will prove interesting, however, to sweep with the glass carefully over the space within
the square, which is comparatively barren to the naked eye,
but in which many small stars will be revealed, of whose existence the naked-eye observer would be unaware. The star
marked Pi (π) is an interesting double, which can be separated by a good eye without artificial aid, and which, with an
opera-glass, presents a fine appearance.

And now we come to Map No. 16, representing the constellations Cetus, Pisces, Aries, and the Triangles. In consulting it the observer is supposed to face the southeast.
Cetus is a very large constellation, and from the peculiar con-
formation of its principal stars it can be readily recognized.
The head is to the east, the star Alpha (α), called Menkar,
being in the nose of this imaginary inhabitant of the sky-
depths. The constellation is supposed to represent the monster that, according to fable, was sent by Neptune to devour
the fair Andromeda, but whose bloodthirsty design was happily and gallantly frustrated by Perseus, as we shall learn
from starry mythology further on.

Although bearing the name Cetus, the Whale, the pictures of the constellation in the old maps do not present us
with the form of a whale, but that of a most extraordinary
scaly creature with enormous jaws filled with large teeth, a
forked tongue, fore-paws armed with gigantic claws, and a
long, crooked, and dangerous-looking tail. Indeed, Aratus
does not call it a "whale," but a "sea-monster," and Dr.
Seiss would have us believe that it was intended to represent

the leviathan, whose terrible prowess is celebrated in the book of Job.

By far the most interesting object in Cetus is the star Mira. This is a famous variable—a sun that sometimes shines a thousand-fold more brilliantly than at others! It changes from the second magnitude to the ninth or tenth, its

MAP 16.

period from maximum to maximum being about eleven months. During about five months of that time it is completely invisible to the naked eye; then it begins to appear again, slowly increasing in brightness for some three months, until it shines as a star of the second magnitude, being then as bright as, if not brighter than, the most brilliant stars in the constellation. It retains this brilliance for about two weeks, and then begins to fade again, and, within three months, once more disappears. There are various irregularities in its changes, which render its exact period somewhat uncertain, and it does not always attain the same degree of brightness at its maximum. For instance, in 1779, Mira was almost equal in brilliance to a first-magnitude star, but frequently at its greatest brightness it is hardly equal to an ordinary star of the second magnitude. By the aid of our little map you will readily be able to find it. You will perceive that it has a slightly reddish tint. Watch it from one of its maxima, and you will see it gradually fade from sight until, at last, only the blackness of the empty sky appears where, a few months before, a conspicuous star was visible. Keep watch of that spot, and in due course you will perceive Mira shining there again—a mere speck, but slowly brightening—and in three months more the wonderful star will blaze again with renewed splendor.

Knowing that our own sun is a variable star—though variable only to a slight degree, its variability being due to the spots that appear upon its surface in a period of about eleven years—we possess some light that may be cast upon the mystery of Mira's variations. It seems not improbable that, in the case of Mira, the surface of the star at the maximum of spottedness is covered to an enormously greater extent than occurs during our own sun-spot maxima, so that the light of the star, instead of being merely dimmed to an almost imperceptible extent, as with our sun, is almost blotted out. When the star blazes with unwonted splendor, as in 1779, we may fairly assume that the pent-up forces of this perishing sun have burst forth, as in a desperate struggle against extinction. But nothing can prevail against the slow, remorseless, unswerving progress of that obscuration, which comes from the leaking away of the solar heat, and which constitutes what we may call the death of a sun. And that word seems peculiarly appropriate to describe the end of a body which, during its period of visible existence, not only presents the highest type of physical activity, but is the parent and supporter of all forms of life upon the planets that surround it.

We might even go so far as to say that possibly Mira presents to us an example of what our sun will be in the course of time, as the dead and barren moon shows us, as in a magician's glass, the approaching fate of the earth. Fortunately, human life is a mere span in comparison with the æons of cosmic existence, and so we need have no fear that either we or our descendants for thousands of generations shall have to play the tragic *rôle* of Campbell's "Last Man," and endeavor to keep up a stout heart amid the crash of time by meanly boasting to the perishing sun, whose rays have nurtured us, that, though his proud race is ended, we have confident anticipations of immortality. I trust that, when man makes his exit from this terrestrial stage, it will not be in the contemptible act of kicking a fallen benefactor.

There are several other variable stars in Cetus, but none possessing much interest for us. The observer should look at the group of stars in the head, where he will find some interesting combinations, and also at Chi, which is the little star shown in the map near Zeta (ζ). This is a double that will serve as a very good test of eye and instrument, the smaller companion-star being of only seven and a half magnitude.

Directly above Cetus is the long, straggling constellation of Pisces, the Fishes. The Northern Fish is represented by the group of stars near Andromeda and the Triangles. A long band or ribbon, supposed to bind the fish together, trends thence first southeast and then west until it joins a group of stars under Pegasus, which represents the Western Fish, not to be confounded with the Southern Fish described near the beginning of this chapter, which is a separate constellation. Fable has, however, somewhat confounded these fishes; for while, as I have remarked above, the Southern Fish is said to represent Venus after she had turned herself into a fish to escape from the giant Typhon, the two fishes of the constellation we are now dealing with are also fabled to represent Venus and her interesting son Cupid under the same disguise assumed on precisely the same occasion. If Typhon, however, was so great a brute that even Cupid's arrows were of no avail against him, we should, perhaps, excuse mythology for duplicating the record of so wondrous an event.

You will find it very interesting to take your glass and, beginning with the attractive little group in the Northern Fish, follow the windings of the ribbon, with its wealth of tiny stars, to the Western Fish. When you have arrived at that point, sweep well over the sky in that neighborhood, and particularly around and under the stars Iota (ι), Theta (θ), Lambda (λ), and Kappa (κ). If you are using a powerful glass, you will be surprised and delighted by what you see. Below the star Omega (ω), and to the left of Lambda, is the place which the sun occupies at the time of the spring equinox—in other words, one of the two crossing-places of the equinoctial or the equator of the heavens, and the ecliptic, or the sun's path. The prime meridian of the heavens passes through this point. You can trace out this great circle, from which astronomical longitudes are reckoned, by drawing an imaginary line from the equinoctial point just indicated through a in ·Andromeda and β in Cassiopeia to the polestar.

To the left of Pisces, and above the head of Cetus, is the constellation Aries, or the Ram. Two pretty bright stars, four degrees apart, one of which has a fainter star near it, mark it out plainly to the eye. These stars are in the head of the Ram. The brightest one, Alpha (a), is called Hamal;

Beta (β) is named Sheratan ; and its fainter neighbor is Me-
sarthim. According to fable, this constellation represents the
ram that wore the golden fleece, which was the object of the
celebrated expedition of the Argonauts. There is not much
in the constellation to interest us, except its historical impor-
tance, as it was more than two thousand years ago the leading
constellation of the zodiac, and still stands first in the list of
the zodiacal signs. Owing to the precession of the equinoxes,
however, the vernal equinoctial point, which was formerly in
this constellation, has now advanced into the constellation
Pisces, as we saw above. Gamma (γ), Arietis, is interesting as
the first telescopic double star ever discovered. Its duplicity
was detected by Dr. Hooke while watching the passage of a
comet near the star in 1664. Singularly enough, the brightest star in the constellation, now bearing the letter α, originally did not belong to the constellation. Tycho Brahe finally
placed it in the head of Aries.

The little constellation of the Triangles, just above Aries,
is worth only a passing notice. Insignificant as it appears,
this little group is a very ancient constellation. It received
its name, Deltoton, from the Greek letter Δ.

The reader must now be introduced to the "Royal Family."
Although the story of Perseus and Andromeda is, of course,
well known to nearly all readers, yet, on account of the great
beauty and brilliancy of the group of constellations that per-
petuate the memory of it among the stars, it is worth recalling here. It will be remembered that, as Perseus was return-
ing through the air from his conquest of the Gorgon Medusa,
he saw the beautiful Andromeda chained to a rock on the
sea-coast, waiting to be devoured by a sea-monster. The
poor girl's only offense was that her mother, Cassiopeia, had
boasted for her that she was fairer than the sea-beauty,
Atergatis, and for this Neptune had decreed that all the
land of the Ethiopians should be drowned and destroyed
unless Andromeda was delivered up as a sacrifice to the

6

dreadful sea-monster.   When Perseus, dropping down to
learn why this maiden was chained to the rocks, heard from
Andromeda's lips the story of her woes, he laughed with

joy. Here was an adventure just to his liking, and besides, unlike his previous adventures, it involved the fate of a beautiful woman with whom he was already in love. Could he save her? Well, wouldn't he! The sea-monster might frighten a kingdom full of Ethiops, but it could not shake the nerves of a hero from Greece. He whispered words of encouragement to Andromeda, who could scarce believe the good news that a champion had come to defend her after all her friends and royal relations had deserted her. Neither could she feel much confidence in her young champion's powers when suddenly her horrified gaze met the awful leviathan of the deep advancing to his feast! But Perseus, with a warning to Andromeda not to look at what he was about to do, sprang with his winged sandals up into the air. And then, as Charles Kingsley has so beautifully told the story—

"On came the great sea-monster, coasting along like a huge black galley, lazily breasting the ripple, and stopping at times by creek or headland to watch for the laughter of girls at their bleaching, or cattle pawing on the sand-hills, or boys bathing on the beach. His great sides were fringed with clustering shells and sea-weeds, and the water gurgled in and out of his wide jaws as he rolled along, dripping and glistening in the beams of the morning sun. At last he saw Andromeda, and shot forward to take his prey, while the waves foamed white behind him, and before him the fish fled leaping.

"Then down from the height of the air fell Perseus like a shooting-star—down to the crest of the waves, while Andromeda hid her face as he shouted. And then there was silence for a while.

"At last she looked up trembling, and saw Perseus springing toward her; and, instead of the monster, a long, black rock, with the sea rippling quietly round it."

Perseus had turned the monster into stone by holding the blood-freezing head of Medusa before his eyes; and it was

fear lest Andromeda herself might see the Gorgon's head, and
suffer the fate of all who looked upon it, that had led him to
forbid her watching him when he attacked her enemy.
Afterward he married her, and Cassiopeia, Andromeda's
mother, and Cepheus, her father, gave their daughter's res-
cuer a royal welcome, and all the Ethiops rose up and blessed
him for ridding the land of the monster. And now, if we
choose, we can, any fair night, see the principal characters of
this old romance shining in starry garb in the sky. Aratus
saw them there in his day, more than two hundred years before Christ, and has left this description in his " Skies," as
translated by Poste :

> " Nor shall blank silence whelm the harassed house
> Of Cepheus ; the high heavens know their name,
> For Zeus is in their line at few removes.
> Cepheus himself by She-bear Cynosure,
> Iasid king stands with uplifted arms.
> From his belt thou castest not a glance
> To see the first spire of the mighty Dragon.

> " Eastward from him, heaven-troubled queen, with scanty stars
> But lustrous in the full-mooned night, sits Cassiopeia.
> Not numerous nor double-rowed
> The gems that deck her form,
> But like a key which through an inward-fastened
> Folding-door men thrust to knock aside the bolts,
> They shine in single zigzag row.
> She, too, o'er narrow shoulders stretching
> Uplifted hands, seems wailing for her child.

> " For there, a woful statue-form, is seen
> Andromeda, parted from her mother's side.   Long I trow
> Thou wilt not seek her in the nightly sky,
> So bright her head, so bright
> Her shoulders, feet, and girdle.
> Yet even there she has her arms extended,
> And shackled even in heaven ; uplifted,
> Outspread eternally are those fair hands.

> " Her feet point to her bridegroom
> Perseus, on whose shoulder they rest.

He in the north-wind stands gigantic,
His right hand stretched toward the throne
Where sits the mother of his bride. As one bent on some high
    deed,
Dust-stained he strides over the floor of heaven."

The makers of old star-maps seem to have vied in the
effort to represent with effect the figures of Andromeda,
Perseus, and Cassiopeia among the stars, and it must be ad-
mitted that some of them succeeded in giving no small degree of life and spirit to their sketches.

The starry riches of these constellations are well matched
with their high mythological repute. Lying in and near the
Milky-Way, they are particularly interesting to the observer
with an opera-glass. Besides, they include several of the
most celebrated wonders of the firmament.

In consulting Map No. 17, the observer is supposed to
face the east and northeast. We will begin our survey with
Andromeda. The three chief stars of this constellation are of
the second magnitude, and lie in a long, bending row, beginning with Alpha (*a*), or Alpheratz, in the head, which, as we
have seen, marks one corner of the great Square of Pegasus.
Beta (*β*), or Mirach, with the smaller stars Mu (*μ*) and Nu (*ν*),
form the girdle. The third of the chief stars is Gamma (*γ*),
or Almaach, situated in the left foot. The little group of
stars designated Lambda (*λ*), Kappa (*κ*), and Iota (*ι*), mark the
extended right hand chained to the rock, and Zeta (*ζ*) and
some smaller stars southwest of it show the left arm and
hand, also stretched forth and shackled.

In searching for picturesque objects in Andromeda, begin
with Alpheratz and the groups forming the hands. Below
the girdle will be seen a rather remarkable arrangement of
small stars in the mouth of the Northern Fish. Now follow
up the line of the girdle to the star Nu (*ν*). If your glass has
a pretty wide field, your eye will immediately catch the glim-
mer of the Great Nebula of Andromeda in the same field

with the star. This is the oldest or earliest discovered of the nebulæ, and, with the exception of that in Orion, is the grandest visible in this hemisphere. Of course, not much can be expected of an opera-glass in viewing such an object ; and yet a good glass, in clear weather and the absence of the moon, makes a very attractive spectacle of it.

By turning the eyes aside, the nebula can be seen, extended as a faint, wispy light, much elongated on either side of the brighter nucleus. The cut here given shows, approximately, the appearance of the nebula, together with some of the small stars in its neighborhood, as seen with a field-glass. With large telescopes it appears both larger and broader, expanding to a truly enormous extent, and in Bond's celebrated picture of it we behold gigantic rifts running lengthwise, while the whole field of sky in which it is contained appears sprinkled over with minute stars apparently between us and the nebula. It was in, or, probably more properly speaking, in line with, this nebula that a new star suddenly shone out in 1885, and, after flickering and fading for a few months, disappeared. That the outburst of light in this star had any real connection with the nebula is exceedingly improbable. Although it appeared to be close beside the bright nucleus of the nebula, it is likely that it was really hundreds or thou-

THE GREAT ANDROMEDA NEBULA.

sands of millions of miles either this side or the other side of
it. Why it should suddenly have blazed into visibility, and
then in so short a time have disappeared, is a question as
difficult as it is interesting. The easiest way to account for
it, if not the most satisfactory, is to assume that it is a variable star of long period, and possessing a very wide range of
variability. One significant fact that would seem to point
to some connection between star and the nebula, after all,
is that a similar occurrence was noticed in the constellation
Scorpio in 1860, and to which I have previously referred (see
Chapter II). In that case a faint star projected against the
background of a nebula, suddenly flamed into comparatively
great brilliance, and then faded again. The chances against
the accidental superposition of a variable star of such extreme variability upon a known nebula occurring twice are so
great that, for that reason alone, we might be justified in
thinking some mysterious causal relation must in each case
exist between the nebula and the star. The temptation to in-
dulge in speculation is very great here, but it is better to
wait for more light, and confess that for the present these
things are inexplicable.

It will be found very interesting to sweep with the glass
slowly from side to side over Andromeda, gradually approaching toward Cassiopeia or Perseus. The increase in the
richness of the stratum of faint stars that apparently forms
the background of the sky will be clearly discernible as you
approach the Milky-Way, which passes directly through
Cassiopeia and Perseus. It may be remarked that the Milky-
Way itself, in that splendidly rich region about Sagittarius
(described in the "Stars of Summer"), is not nearly so effective an object with an opera-glass as it is above Cygnus and in
the region with which we are now dealing. This seems to be
owing to the smaller magnitude of its component stars in the
southern part of the stream. There the background appears
more truly "milky," while in the northern region the little

stars shine distinct, like diamond-specks, on a black background.

The star Nu, which serves as a pointer to the Great Nebula, is itself worth some attention with a pretty strong glass on account of a pair of small stars near it.

The star Gamma (γ) is interesting, not only as one of the most beautiful triples in the heavens (an opera-glass is far too feeble an instrument to reveal its companions), but because it serves to indicate the radiant point of the Biela meteors. There was once a comet well known to astronomers by the name of its discoverer, Biela. It repeated its visits to the neighborhood of the sun once in every six or seven years. In 1846 this comet astonished all observers by splitting into two comets, which continued to run side by side, like two equal racers, in their course around the sun. Each developed a tail of its own. In 1852, when the twin comets were due again, the astronomical world was on the *qui vive*, and they did not disappoint expectation, for back they came out of the depths of space, still racing, but much farther apart than they had been before, alternating in brightness as if the long struggle had nearly exhausted them, and finally, like spent runners, growing faint and disappearing. They have never been seen since.

In 1872, when the comets should have been visible, if they still existed, a very startling thing happened. Out of the northern heavens, along the track of the missing comets, where the earth crossed it, on the night of the 27th of November came glistening and dashing the fiery spray of a storm of meteors. It was the dust and fragments of the lost comet of Biela, which, after being split in two in 1852, had evidently continued the process of disintegration until its cometary character was completely lost. It seems to have made a truly ghostly exit, for right after the meteor swarm of 1872 a mysterious cometary body was seen, which was supposed at the time to be the missing comet itself, and which, it is not alto-

gether improbable, may have been a fragment of it. Three
days after the meteors burst over Europe, it occurred to Pro-
fessor Klinkerfues, of Berlin, that if they came from Biela's
comet the comet itself ought to be seen in the southern
hemisphere retreating from its encounter with the earth. On
November 30th he sent his now historical telegram to Mr.
Pogson, an astronomer at Madras; "Biela touched earth November 27th. Search near Theta Centauri." For thirty-six
hours after the receipt of this extraordinary request Mr. Pogson was prevented by clouds from scanning the heavens with
his telescope. When the sky cleared at last, behold there
was a comet in the place indicated in the telegram! It was
glimpsed again the next night, and then clouds intervened,
and not a trace of it was ever seen afterward.

But every year, on the 27th of November, when the earth
crosses the orbit of the lost comet, meteoric fragments come
plunging into our atmosphere, burning as they fly. Ordinarily their number is small, but when, as in 1872, a swarm of
the meteors is in that part of their orbit which the earth
crosses, there is a brilliant spectacle. In 1885 this occurred,
and the world was treated to one of the most splendid meteoric displays on record.

Next let us turn to Perseus. The bending row of stars
marking the center of this constellation is very striking and
brilliant. The brightest star in the constellation is Alpha, or
Algenib, in the center of the row. The head of Perseus is
toward Cassiopeia, and in his left hand he grasps the head of
Medusa, which hangs down in such a way that its principal
star Beta, or Algol, forms a right angle with Algenib and
Almaach in Andromeda. This star Algol, or the Demon, as
the Arabs call it, is in some respects the most wonderful and
interesting in all the heavens. It is as famous for the variability of its light as Mira, but it differs widely from that star
both in its period, which is very short, and in the extent of
the changes it undergoes. During about two days and a

half, Algol is equal in brilliance to Algenib, which is a second-magnitude star; then it begins to fade, and in the course of about four and a half hours it sinks to the fourth magni-

THE ATTENDANTS OF ALPHA PERSEI.

tude, being then about equal to the faint stars near it. It remains thus obscured for only a few minutes, and then begins to brighten again, and in about four and a half hours more resumes its former brilliance. This phenomenon is very easily observed, for, as will be seen by consulting our little map, Algol can be readily found, and its changes are so rapid that under favorable circumstances it can be seen in the course of a single night to run through the whole gamut.

Of course, no optical instrument whatever is needed to enable one to see these changes of Algol, for it is plainly visible to the naked eye throughout, but it will be found interesting to watch the star with an opera-glass. Its periodic time from minimum to minimum is two days, twenty hours, and forty-nine minutes, lacking a few seconds. Any one can calculate future minima for himself by adding the periodic time above given to the time of any observed minimum.

While spots upon its surface may be the cause of the variations in the light of Mira, it is believed that the more rapid changes of Algol may be due to another cause ; namely, the existence of a huge, dark body revolving swiftly around it at close quarters in an orbit whose plane is directed edge-wise toward the earth, so that at regular intervals this dark body causes a partial eclipse of Algol. Notwithstanding the attacks that have been made upon this theory, it seems to hold its ground, and it will probably continue to find favor as a working hypothesis until some fresh light is cast upon the problem. It hardly needs to be said that the dark body in question, if it exists, must be of enormous size, bearing no such insignificant proportion to the size of Algol as the earth does to the sun, but being rather the rival in bulk of its shining brother—a blind companion, an extinguished sun.

There was certainly great fitness in the selection of the little group of stars of which this mysterious Algol forms the most conspicuous member, to represent the awful head of the Gorgon carried by the victorious Perseus for the confusion of his enemies. In a darker age than ours the winking of this demon-star must have seemed a prodigy of sinister import.

Turn now to the bright star Algenib, or Alpha Persei. You will find with the glass an exceedingly attractive spec-tacle there. In my note-book I find this entry, made while sweeping over Perseus for materials for this chapter: "The field about Alpha is one of the finest in the sky for an opera-glass. Stars conspicuously ranged in curving lines and

streams. A host follows Alpha from the east and south."
The picture on page 84 will give the reader some notion of
the exceeding beauty of this field of stars, and of the singular
manner in which they are grouped, as it were, behind their
leader. A field-glass increases the beauty of the scene.

The reader will find a starry cluster marked on Map 17 as
the "Great Cluster." This object can be easily detected by
the naked eye, resembling a wisp of luminous cloud. It
marks the hand in which Perseus clasps his diamond sword,
and, with a telescope of medium power, it is one of the most
marvelously beautiful objects in the sky—a double swarm of
stars, bright enough to be clearly distinguished from one another, and yet so numerous as to dazzle the eye with their
lively beams. An opera-glass does not possess sufficient
power to "resolve" this cluster, but it gives a startling suggestion of its half-hidden magnificence, and the observer will
be likely to turn to it again and again with increasing admiration. Sweep from this to Alpha Persei and beyond to get
an idea of the procession of suns in the Milky-Way. The
nebulous-looking cluster marked 34 M appears with an opera-
glass like a faint comet.

About a thousand years ago the theologians undertook to
reconstruct the constellation figures, and to give them a re-
ligious significance. They divided the zodiac up among the
twelve apostles, St. Peter taking the place of Aries, with the
Triangles for his mitre. In this reconstruction Perseus was
transmogrified into St. Paul, armed with a sword in one hand
and a book in the other; Cassiopeia became Mary Magda-
lene; while poor Andromeda, stripped of all her beauty and
romance, was turned into a sepulchre!

Next look at Cassiopeia, which is distinctly marked out
by the zigzag row of stars so well described by Aratus. Here
the Milky-Way is so rich that the observer hardly needs any
guidance; he is sure to stumble upon interesting sights for
himself. The five brightest stars are generally represented as

indicating the outlines of the chair or throne in which the queen sits, the star Zeta (ζ) being in her head. Look at Zeta with a good field-glass, and you will see a singular and brilliant array of stars near it in a broken half-circle, which may suggest the notion of a crown. Near the little star Kappa (κ) in the map will be seen a small circle and the figures 1572. This shows the spot where the famous temporary star, which has of late been frequently referred to as the "Star of Bethlehem," appeared. It was seen in 1572, and carefully observed by the famous astronomer Tycho Brahe. It seems to have suddenly burst forth with a brilliance that outshone every other star in the heavens, not excepting Sirius itself. But its supremacy was short-lived. In a few months it had sunk to the second magnitude. It continued to grow fainter, exhibiting some remarkable changes of color in the mean time, and in less than a year and a half it disappeared. It has never been seen since. But in 1264, and again in 945, a star is said to have suddenly blazed out near that point in the heavens. There is no certainty about these earlier apparitions, but, assuming that they are not apocryphal, they might possibly indicate that the star seen by Tycho was a periodical one, its period considerably exceeding three hundred years. Carrying this supposed period back, it was found that an apparition of this star might have occurred about the time of the birth of Christ. It did not require a very prolific imagination to suggest its identity with the so-called star of the Magi, and hence the legend of the Star of Bethlehem and its impending reappearance, of which we have heard so much of late. It will be observed, from the dates given above, that, even supposing them to be correct, no definite period is indicated for the reappearance of the star. In one case the interval is three hundred and eight years, and in the other three hundred and nineteen years. In short, there are too many suppositions and assumptions involved to allow of any credence being given to the theory of the periodicity of Tycho's wonderful star. At the same time, nobody can say it is impossible that the star should appear again, and so it may be interesting for the reader to know where to look for it.

Many of the most beautiful sights of this splendid constellation are beyond the reach of an opera-glass, and reserved for the grander powers of the telescope.

We will pause but briefly with Cepheus, for the old king's constellation is comparatively dim in the heavens, as his part in the dramatic story of Andromeda was contemptible, and he seems to have got among the stars only by virtue of his relationship to more interesting persons. He does possess one gem of singular beauty—the star Mu, which may be found about two and a half degrees south of the star Nu (ν). It is the so-called "Garnet Star," thus named by William Herschel, who advises the observer, in order to appreciate its color, to glance from it to Alpha Cephei, which is a white star. Mu is variable, changing from the fourth to the sixth magnitude in a long period of five or six years. Its color is changeable, like its light. Sometimes it is of a deep garnet hue, and at other times it is orange-colored. Upon the whole, it appears of a deeper red than any other star visible to the naked eye.

If you have a good field-glass, try its powers upon the star Delta (δ) Cephei. This is a double star, the components being about forty-one seconds of arc apart, the larger of four and one half magnitude, and the smaller of the seventh magnitude. The latter is of a beautiful blue color, while the larger star is yellow or orange. With a good eye, a steady hand, and a clear glass, magnifying not less than six diameters, you can separate them, and catch the contrasted tints of their light. Besides being a double star, Delta is variable.

CHAPTER IV.

I HAVE never beheld the first indications of the rising of Orion without a peculiar feeling of awakened expectation, like that of one who sees the curtain rise upon a drama of absorbing interest. And certainly the magnificent company of the winter constellations, of which Orion is the chief, make their entrance upon the scene in a manner that may be described as almost dramatic. First in the east come the world-renowned Pleiades. At about the same time Capella, one of the most beautiful of stars, is seen flashing above the north-eastern horizon. These are the sparkling ushers to the coming spectacle. In an hour the fiery gleam of Aldebaran appears at the edge of the dome below the Pleiades, a star noticeable among a thousand for its color alone, besides being one of the brightest of the heavenly host. The observer familiar with the constellations knows, when he sees this red star which marks the eye of the angry bull, Taurus, that just behind the horizon stands Orion with starry shield and up-raised club to meet the charge of his gigantic enemy. With Aldebaran rises the beautiful V-shaped group of the Hyades. Presently the star-streams of Eridanus begin to appear in the east and southeast, the immediate precursors of the rising of Orion :

> " And now the river-flood's first winding reach
> The becalmed mariner may see in heaven,
> As he watches for Orion to espy if he hath aught to say
> Of the night's measure or the slumbering winds."

The first glimpse we get of the hero of the sky is the long bending row of little stars that glitter in the lion's skin which, according to mythology, serves him for a shield. The great constellation then advances majestically into sight. First of its principal stars appears Bellatrix in the left shoulder; then the little group forming the head, followed closely by the splendid Betelgeuse, "the martial star," flashing like a decoration upon the hero's right shoulder. Then come into view the equally beautiful Rigel in the left foot, and the striking row of three bright stars forming the Belt. Below these hangs another starry pendant marking the famous sword of Orion, and last of all appears Saiph in the right knee. There is no other constellation containing so many bright stars. It has two of the first magnitude, Betelgeuse and Rigel; the three stars in the Belt, and Bellatrix in the left shoulder, are all of the second magnitude; and besides these there are three stars of the third magnitude, more than a dozen of the fourth, and innumerable twinklers of smaller magnitudes, whose commingled scintillations form a celestial illumination of singular splendor.

"Thus graced and armed he leads the starry host."

By the time Orion has chased the Bull half-way up the eastern slope of the firmament, the peerless Dog-Star, Sirius, is flaming at the edge of the horizon, while farther north glitters Procyon, the little Dog-Star, and still higher are seen the twin stars in Gemini. When these constellations have advanced well toward the meridian, as shown in our circular map, their united radiance forms a scene never to be forgotten. Counting one of the stars in Gemini as of the first rank, there are no less than seven first-magnitude stars ranged around one another in a way that can not fail to attract the attention and the admiration of the most careless observer. Aldebaran, Capella, the Twins, Procyon, Sirius, and Rigel mark the angles of a huge hexagon, while Betelgeuse shines with ruddy beauty not far from the center of the figure. The heavens contain no other naked-eye view comparable with this great array, not even the glorious celestial region where the Southern Cross shines supreme, being equal to it in splendor.

As an offset to the discomforts of winter observations of the stars, the observer finds that the softer skies of summer have no such marvelous brilliants to dazzle his eyes as those that illumine the hyemal heavens. To comprehend the real glories of the celestial sphere in the depth of winter one should spend a few clear nights in the rural districts of New York or New England, when the hills, clad with sparkling blankets of crusted snow, reflect the glitter of the living sky. In the pure frosty air the stars seem splintered and multiplied indefinitely, and the brighter ones shine with a splendor of light and color unknown to the denizen of the smoky city, whose eyes are dulled and blinded by the glare of street-lights. There one may detect the delicate shade of green that lurks in the imperial blaze of Sirius, the beautiful rose-red light of Aldebaran, the rich orange hue of Betelguese, the blue-white radiance of Rigel, and the pearly luster of Capella. If you have never seen the starry heavens except as they appear from city streets and squares, then, I had almost said, you have never seen them at all, and especially in the winter is this true. I wish I could describe to you the impression that they can make upon the opening mind of a country boy, who, knowing as yet nothing of the little great world around him, stands in the yawning silence of night and beholds the illimitably great world above him, looking deeper than thought can go into the shining vistas of the universe, and overwhelmed with the wonder of those marshaled suns.

Looking now at Map 18, we see the heavens as they appear at midnight on the 1st of December, at 10 o'clock P. M. on the 1st of January, and at 8 o'clock P. M. on the 1st of

7

February.   In the western half of the sky we recognize Andromeda, Pegasus, Pisces, Cetus, Aries, Cassiopeia, and other

MAP. 18.

constellations that we studied in the "Stars of Autumn."
Far over in the east we see rising Leo, Cancer, and Hydra,
which we included among the "Stars of Spring." Occupying
most of the southern and eastern heavens are the constellations which we are now to describe under the name of the

"Stars of Winter," because in that season they are seen under the most favorable circumstances. I have already referred to the admirable way in which the principal stars of some of these constellations are ranged round one another.

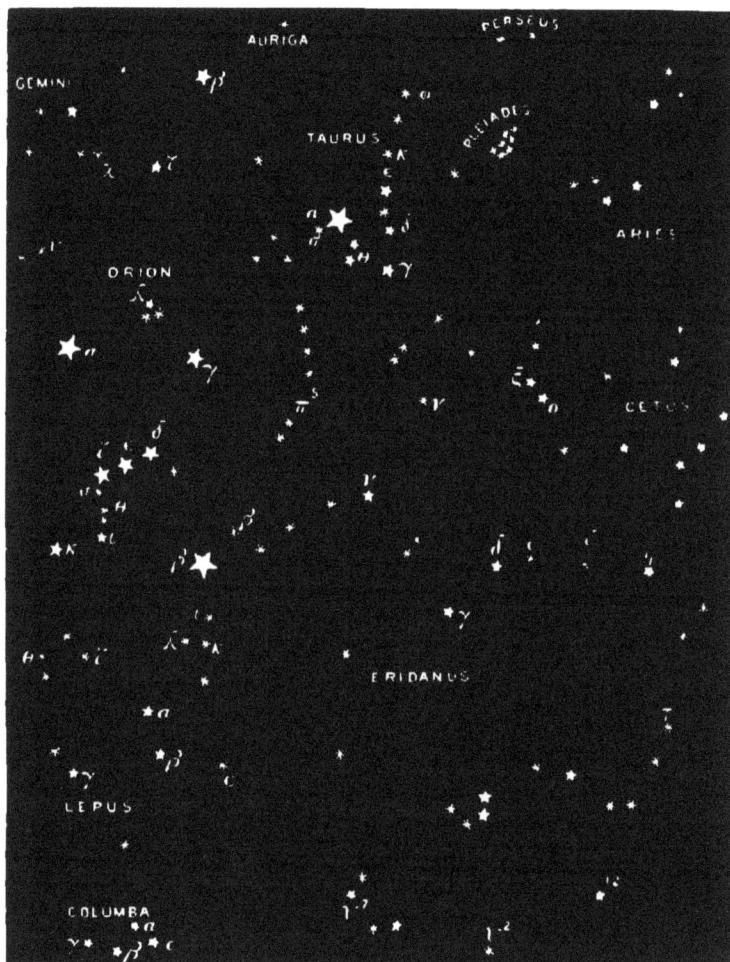

MAP 19.

By the aid of the map the observer can perceive the relative position of the different constellations, and, having fixed this in his mind, he will be prepared to study them in detail.

Let us now begin with Map No. 19, which shows us the constellations of Eridanus, Lepus, Orion, and Taurus. Eridanus is a large though not very conspicuous constellation, which is generally supposed to represent the celebrated river now known as the Po. It has had different names among different peoples, but the idea of a river, suggested by its long, winding streams of stars, has always been preserved. According to fable, it is the river into which Phaeton fell after his disastrous attempt to drive the chariot of the sun for his father Phœbus, and in which hare-brained adventure he narrowly missed burning the world up. The imaginary river starts from the brilliant star Rigel, in the left foot of Orion, and flows in a broad upward bend toward the west; then it turns in a southerly direction until it reaches the bright star Gamma (γ), where it bends sharply to the north, and then quickly sweeps off to the west once more, until it meets the group of stars marking the head of Cetus. Thence it runs south, gradually turning eastward, until it flows back more than half-way to Orion. Finally it curves south again and disappears beneath the horizon. Throughout the whole distance of more than 100° the course of the stream is marked by rows of stars, and can be recognized without difficulty by the amateur observer.

The first thing to do with your opera-glass, after you have fixed the general outlines of the constellation in your mind by naked-eye observations, is to sweep slowly over the whole course of the stream, beginning at Rigel, and following its various wanderings. Eridanus ends in the southern hemisphere near a first-magnitude star called Achernar, which is situated in the stream, but can not be seen from our latitudes. Along the stream you will find many interesting groupings of the stars. In the map see the pair of stars below and to the right of Nu (ν). These are the two Omicrons, the upper one being

o^1 and the lower one o^2. The latter is of an orange hue, and is remarkable for the speed with which it is flying through space. There are only one or two stars whose proper motion, as it is called, is more rapid than that of o^2 in Eridanus. It changes its place nearly seven minutes of arc in a century. The records of the earliest observations we possess show that near the beginning of the Christian era it was about half-way between o^1 and ν. Its companion o^1, on the contrary, seems to be almost stationary, so that o^2 will gradually draw away from it, passing on toward the southwest until, in the course of centuries, it will become invisible from our latitudes. This flying star is accompanied by two minute companions, which in themselves form a close and very delicate double star. These two little stars, of only 9·5 and 10·5 magnitude, respectively, are, of course beyond the ken of the observer with an opera-glass. The system of which they form a part, however, is intensely interesting, since the appearances indicate that they belong, in the manner of satellites, to o^2, and are fellow-voyagers of that wonderful star.

Having admired the star-groups of Eridanus, one of the prettiest of which is to be seen around Beta (β), let us turn next to Taurus, just above or north of Eridanus. Two remarkable clusters at once attract the eye, the Hyades, which are shaped somewhat like the letter V, with Aldebaran in the upper end of the left-hand branch, and the Pleiades, whose silvery glittering has made them celebrated in all ages. The Pleiades are in the shoulder and the Hyades in the face of Taurus, Aldebaran most appropriately representing one of his blazing eyes as he hurls himself against Orion. The constellation-makers did not trouble themselves to make a complete Bull, and only the head and fore-quarters of the animal are represented. If Taurus had been completed on the scale on which he was begun, there would have been no room in the sky for Aries ; one of the Fishes would have had to abandon his celestial swimming-place, and even the fair Andromeda

would have found herself uncomfortably situated. But, as if
to make amends for neglecting to furnish their heavenly Bull
with hind-quarters, the ancients gave him a most prodigious
and beautiful pair of horns, which make the beholder feel
alarm for the safety of Orion. Starting out of the head
above the Hyades, as illustrated in our cut, the horns curve
upward and to the east, each being tipped by a bright star.
Along and between the horns runs a scattered and broken

THE "GOLDEN HORNS" OF TAURUS.

stream of minute stars which seem to be gathered into knots
just beyond the end of the horns, where they dip into the
edge of the Milky-Way. Many of these stars can be seen, on
a dark night, with an ordinary opera-glass, but, to see them
well, one should use as large a field-glass as he can obtain.
With such a glass their appearance almost makes one suspect
that Virgil had a poetic prevision of the wonders yet to be
revealed by the telescope when he wrote, as rendered by Dry-
den, of the season—

> " When with his *golden horns* in full career
> The Bull beats down the barriers of the year."

Below the tips of the horns, and over Orion's head, there are also rich clusters of stars, as if the Bull were flaunting shreds of sparkling raiment torn from some celestial victim of his fury. With an ordinary glass, however, the observer will not find this star-sprinkled region around the horns of Taurus as brilliant a spectacle as that presented by the Hyades and the group of stars just above them in the Bull's ear. The two stars in the tips of the horns are both interesting, each in a different way. The upper and brighter one of the two, marked Beta (β) in Map No. 19, is called El Nath. It is common to the left horn of Taurus and the right foot of Auriga, who is represented standing just above. It is a singularly white star. This quality of its light becomes conspicuous when it is looked at with a glass. The most inexperienced observer will hardly fail to be impressed by the pure whiteness of El Nath, in comparison with which he will find that many of the stars he had supposed to be white show a decided tinge of color. The star in the tip of the right or southern horn, Zeta (ζ), is remarkable, not on its own account, but because it serves as a pointer to a famous nebula, the discovery of which led Messier to form his catalogue of nebulæ. This is sometimes called the "Crab Nebula," from the long sprays of nebulous matter which were seen surrounding it with Lord Rosse's great telescope. Our little sketch is simply intended to enable the observer to locate this strange object. If he wishes to study its appearance, he must use a powerful telescope. But with a first-rate field-glass he can see it as a speck of light in the position shown in the cut, where the large star is Zeta and the smaller ones are faint stars, the relative position of which will enable the observer to find the nebula, if he keeps in mind that the top of the cut is toward the north. It is noteworthy that this nebula for a time deceived several of the watchers who were on the lookout for the predicted return of Halley's comet in 1835.

And now let us look at the Hyades, an assemblage of stars not less beautiful than their more celebrated sisters the Pleiades. The leader of the Hyades is Aldebaran, or Alpha Tauri, and his followers are worthy of their leader. The inexperienced observer is certain to be surprised by the display of stars which an opera-glass brings to view in the Hyades. Our illustration will give some notion of their appearance with a large field-glass. The "brackish poet," of whose rhymes Admiral Smyth was so fond, thus describes the Hyades:

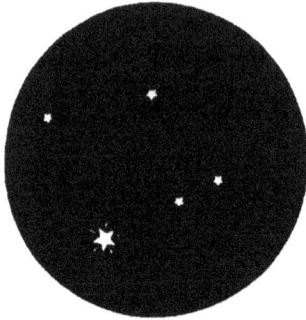

THE CRAB NEBULA.

"In lustrous dignity aloft see Alpha Tauri shine,
The splendid zone he decorates attests the Power divine :
For mark around what glitt'ring orbs attract the wandering eye,
You'll soon confess no other star has such attendants nigh."

The redness of the light of Aldebaran is a very interesting phenomenon. Careful observation detects a decided difference between its color and that of Betelgeuse, or Alpha Orionis, which is also a red star. It differs, too, from the brilliant red star of summer, Antares.. Aldebaran has a trace of rose-color in its light, while Betelgeuse is of a very deep orange, and Antares may be described as fire-red. These shades of color can easily be detected by the naked eye after a little practice. First compare Aldebaran and Betelgeuse, and glance from each to the brilliant white, or bluish-white, star Rigel in Orion's foot. Upon turning the eye back from Rigel to Aldebaran the peculiar color of the latter is readily perceived. Spectroscopic analysis has revealed the presence in Aldebaran of hydrogen, sodium, magnesium, calcium, iron, bismuth, tellurium, antimony, and mercury. And so modern discoveries, while they have pushed back the stars to distances of which the ancients could not conceive, have, at the same time, and equally, widened the recognized boundaries of the physical universe and abolished forever the ancient distinction between the heavens and the earth. It is a plain road from the earth to the stars, though mortal feet can not tread it.

Keeping in mind that in our little picture of the Hyades the top is north, the right hand west, and the left hand east, the reader will be able to identify the principal stars in the group. Aldebaran is readily recognized, because it is the largest of all. The bright star near the upper edge of the picture is Epsilon Tauri, and its sister star, forming the point of the V, is Gamma Tauri. The three brightest stars between Epsilon and Gamma, forming a little group, are the Deltas, while the pair of stars surrounded by many smaller ones, half-way between Aldebaran and Gamma, are the Thetas. These stars present a very pretty appearance, viewed with a good glass, the effect being heightened by a contrast of color in the two Thetas.

The little pair southeast of Aldebaran, called the Sigmas, is also a beautiful object. The distance apart of these stars is about seven minutes of arc, while the distance between the two Thetas is about five and a half minutes of arc. These measures may be useful to the reader in estimating the distances between other stars that he

THE HYADES.

may observe.   It will also be found an interesting test of
the eye-sight to endeavor to see these stars as doubles without the aid of a glass.   Persons having keen eyes will be
able to accomplish this.

North of the star Epsilon will be seen a little group in the
ear of the Bull (see cut, "The Golden Horns of Taurus"),
which presents a brilliant appearance with a small glass.
The southernmost pair in the group are the Kappas, whose
distance apart is very nearly the same as that of the Thetas,
described above ; but I think it improbable that anybody
could separate them with the naked eye, as there is a full
magnitude between them in brightness, and the smaller star
is only of magnitude 6·5, while sixth-magnitude stars are
generally reckoned as the smallest that can be seen by the
naked eye.   Above the Kappas, and in the same group in the
ear, are the two Upsilons, forming a wider pair.

Next we come to the Pleiades :

" Though small their size and pale their light, wide is their fame."

In every age and in every country the Pleiades have been
watched, admired, and wondered at, for they are visible from
every inhabited land on the globe.   To many they are popularly known as the Seven Stars, although few persons can see
more than six stars in the group with the unaided eye.   It is
a singular fact that many of the earliest writers declare that
only six Pleiades can be seen, although they all assert that
they are seven in number.   These seven were the fabled
daughters of Atlas, or the Atlantides, whose names were
Merope, Alcyone, Celæno, Electra, Taygeta, Asterope, and
Maia.   One of the stories connected with them is that Merope
married a mortal, whereupon her star grew dim among her
sisters.   Another fable assures us that Electra, unable to en-
dure the sight of the burning of Troy, hid her face in her
hands, and so blotted her star from the sky.   While we may
smile at these stories, we can not entirely disregard them, for

they are intermingled with some of the richest literary treas-
ures of the world, and they come to us, like some old keep-
sake, perfumed with the memory of a past age. The mythological history of the Pleiades is intensely interesting, too,
because it is world-wide. They have impressed their mark,
in one way or another, upon the habits, customs, traditions,
language, and history of probably every nation. This is true
of savage tribes as well as of great empires. The Pleiades
furnish one of the principal links that appear to connect the
beginnings of human history with that wonderful prehistoric
past, where, as through a gulf of mist, we seem to perceive
faintly the glow of a golden age beyond. The connection of
the Pleiades with traditions of the Flood is most remarkable.
In almost every part of the world, and in various ages, the
celebration of a feast or festival of the dead, dimly connected
by traditions with some great calamity to the human race in
the past, has been found to be directly related to the Pleiades. This festival or rite, which has been discovered in
various forms among the ancient Hindoos, Egyptians, Per-
sians, Peruvians, Mexicans, Druids, etc., occurs always in
the month of November, and is regulated by the culmination
of the Pleiades. The Egyptians directly connected this celebration with a deluge, and the Mexicans, at the time of the
Spanish conquest, had a tradition that the world had once
been destroyed at the time of the midnight culmination of the
Pleiades. Among the savages inhabiting Australia and the
Pacific island groups a similar rite has been discovered. It
has also been suggested that the Japanese feast of lanterns is
not improbably related to this world-wide observance of the
Pleiades, as commemorating some calamitous event in the far
past which involved the whole race of man in its effects.

The Pleiades also have a supposed connection with that
mystery of mysteries, the great Pyramid of Cheops. It has
been found that about the year 2170 B. C., when the beginning of spring coincided with the culmination of the Pleiades

at midnight, that wonderful group of stars was visible, just at midnight, through the mysterious southward-pointing passage of the Pyramid. At the same date the then pole-star, Alpha Draconis, was visible through the northward-pointing passage of the Pyramid.

Another curious myth involving the Pleiades as a part of the constellation Taurus is that which represents this constellation as the Bull into which Jupiter changed himself when he carried the fair Europa away from Phœnicia to the continent that now bears her name. In this story the fact that only the head and fore-quarters of the Bull are visible in the sky is accounted for on the ground that the remainder of his body is beneath the water through which he is swimming. Here, then, is another apparent link with the legends of the Flood, with which the Pleiades have been so strangely connected, as by common consent among many nations, and in the most widely separated parts of the earth.

With the most powerful field-glass you may be able to see all of the stars represented in our picture of the Pleiades. With an ordinary opera-glass the fainter ones will not be visible; yet even with such a glass the scene is a remarkable one. Not only all of the "Seven Sisters," but many other stars, can be seen twinkling among them. The superiority of Alcyone to the others, which is not so clear to the naked eye, becomes very apparent. Alcyone is the large star below the middle of the picture with a triangle of little stars beside it. To the left or east of Alcyone the two most conspicuous stars are Atlas and Pleione. The latter—which is the uppermost one—is represented too large in the picture. It requires a sharp eye to see Pleione without a glass, while Atlas is plainly visible to the unaided vision, and is always counted among the naked-eye Pleiades, although it does not bear the name of one of the mythological sisters, but that of their father. The bright star below and to the right of Alcyone is Merope; the one near the right-hand edge of the picture, about on a

level with Alcyone, is Electra. Above, or to the north of
Electra, are two
bright stars lying
in a line pointing
toward Alcyone;
the upper one of
these, or the one
farthest from Al-
cyone, is Taygeta,
and the other is
Maia. Above Tay-
geta and Maia, and
forming a little tri-
angle with them, is
a pair of stars which
bears the name of

THE PLEIADES.

Asterope. About half-way between Taygeta and Electra, and
directly above the latter, is Celæno.

The naked-eye observer will probably find it difficult
to decide which he can detect the more easily, Celæno or
Pleione, while he will discover that Asterope, although composed of two stars, as seen with a glass, is so faint as to be
much more difficult than either Celæno or Pleione. Unless,
as is not improbable, the names have become interchanged
in the course of centuries, the brightness of these stars would
seem to have undergone remarkable changes. The star of
Merope, it will be remembered, was said to have become in-
distinct, or disappeared, because she married a mortal. At
present Merope is one of those that can be plainly seen with
the naked-eye, while the star of Asterope, who was said to
have had the god Mars for her spouse, has faded away until
only a glass can show it. It would appear, then, that notwithstanding an occasional temporary eclipse, it is, in the
long run, better to marry a plain mortal than a god. Electra,
too, who hid her eyes at the sight of burning Troy, seems to

have recovered from her fright, and is at present, next to
Alcyone, the brightest star in the cluster. But, however we
may regard those changes in the brightness of the Pleiades
which are based upon tradition, there is no doubt that well-
attested changes have taken place in the comparative brilliancy of stars in this cluster since astronomy became an exact science.

Observations of the proper motions of the Pleiades have
shown that there is an actual physical connection between
them; that they are, literally speaking, a flight of suns.
Their common motion is toward the southwest, under the
impulse of forces that remain as yet beyond the grasp of
human knowledge. Alcyone was selected by Mädler as the
central sun around which the whole starry system revolved,
but later investigations have shown that his speculation was
not well founded, and that, so far as we can determine, the
proper motions of the stars are not such as to indicate the
existence of any common center. They appear to be flying
with different velocities in every direction, although—as in
the case of the Pleiades—we often find groups of them associated together in a common direction of flight.

Still another curious fact about the Pleiades is the existence of some rather mysterious nebulous masses in the cluster. In 1859 Temple discovered an extensive nebula, of a
broad oval form, with the star Merope immersed in one end
of it. Subsequent observations showed that this strange
phenomenon was variable. Sometimes it could not be seen;
at other times it was very plain and large. In Jeaurat's
chart of the Pleiades, made in 1779, a vast nebulous mass
is represented near the stars Atlas and Pleione. This has
since been identified by Goldschmidt as part of a huge, ill-
defined nebula, which he thought he could perceive enveloping the whole group of the Pleiades. Many observers,
however, could never see these nebulous masses, and were
inclined to doubt their actual existence. Within the past

few years astronomical photography, having made astonishing progress, has thrown new light upon this mysterious subject. The sensitized plate of the camera, when applied at the focus of a properly constructed telescope, has proved more effective than the human retina, and has, so to speak, enabled us to see beyond the reach of vision by means of the pictures it makes of objects which escape the eye. In November, 1885, Paul and Prosper Henry turned their great photographing telescope upon the Pleiades, and with it discovered a nebula apparently attached to the star Maia. The most powerful telescopes in the world had never revealed this to the eye. Yet of its actual existence there can be no question. Their photograph also showed the Merope nebula, although much smaller, and of a different form from that represented by its discoverer and others. There evidently yet remains much to be discovered in this singular group, and the mingling of nebulous matter with its stars makes Tennyson's picturesque description of the Pleiades appear all the more life-like :

"Many a night I saw the Pleiads, rising through the mellow shade,
Glitter like *a swarm of fire-flies tangled in a silver braid.*"

The reader should not expect to be able to see the nebulæ in the Pleiades with an opera-glass. I have thought it proper to mention these singular objects only in order that he might be in possession of the principal and most curious facts about those interesting stars.*

* The Henry Brothers have continued the photographic work described above, and their later achievements are even more interesting and wonderful. They have found that there are many nebulous masses involved in the group of the Pleiades, and have photographed them. One of the most amazing phenomena in their great photograph of the Pleiades is a long wisp or streak of nebulous matter, along which eight or nine stars are strung in a manner which irresistibly suggests an intimate connection between the stars and the nebula. This recalls the recent (August, 1888) discovery made by Prof. Holden, with the great Lick telescope, concerning the structure of the celebrated ring nebula in Lyra, which, it appears, is composed of concentric ovals of stars and nebulous stuff, so arranged that we must believe they are intimately associated in a most wonderful community.

Orion will next command our attention. You will find the constellation in Map No. 19 :

" Eastward beyond the region of the Bull
Stands great Orion ; whoso kens not him in cloudless night
Gleaming aloft, shall cast his eyes in vain
To find a brighter sign in all the heaven."

To the naked eye, to the opera-glass, and to the telescope, Orion is alike a mine of wonders. This great constellation embraces almost every variety of interesting phenomena that the heavens contain. Here we have the grandest of the nebulæ, some of the largest and most beautifully colored stars, star-streams, star-clusters, nebulous stars, variable stars. I have already mentioned the positions of the principal stars in the imaginary figure of the great hunter. I may add that his upraised arm and club are represented by the stars seen in the map above Alpha (*a*) or Betelgeuse, one of which is marked Nu (*v*), and another, in the knob of the club, Chi (*χ*). I have also, in speaking of Aldebaran, described the contrast in the colors of Betelgeuse and Beta (*β*) or Rigel. Betelgeuse, it may be remarked, is slightly variable. Sometimes it appears brighter than Rigel, and sometimes less brilliant. It is interesting to note that, according to Secchi's division of the stars into types, based upon their spectra, Betelgeuse falls into the third order, which seems to represent a type of suns in which the process of cooling, and the formation of an absorptive envelope or shell, have gone on so far that we may regard them as approaching the point of extinction. Rigel, on the other hand, belongs to the first order or type which represents suns that are probably both hotter and younger in the order of development. So, then, we may look upon the two chief stars of this great constellation as representing two stages of cosmical existence. Betelgeuse shows us a sun that has almost run its course, that has passed into its decline, and that already begins to faint and flicker and grow dim before the on-coming and inevitable fate of extinction ; but in Rigel we see a sun blazing with the fires of youth, splendid in the first glow of its solar energies, and holding the promise of the future yet before it. Rigel belongs to a new generation of the universe; Betelgeuse to the universe that is passing. We may pursue this comparison one step farther back and see in the great nebula, which glows dimly in the middle of the constellation, between Rigel triumphant and Betelgeuse languishing, a still earlier cosmical condition— the germ of suns whose infant rays may illuminate space when Rigel itself is growing dim.

Turn your glass upon the three stars forming the Belt. You will not be likely to undertake to count all the twinkling lights that you

THE SWORD OF ORION AND THE GREAT NEBULA.

will see, especially as many of them appear and disappear as you turn your attention to different parts of the field. Sweep all around the Belt and also between the Belt and Gamma (γ) or Bellatrix. According to the old astrologers, women born under the influence of the star Bellatrix were lucky, and provided with good tongues. Of course, this was fortunate for their husbands too!

Below the Belt will be seen a short row of stars hanging downward and representing the sword. In the middle of this row is the great Orion nebula. The star Theta (θ) involved in the nebula is multiple, and the position of this little cluster of suns is such that, as has been said, they seem to be feeding upon the substance of the nebula surrounding them. Other stars are seen scattered in different parts of the nebula. This phenomenon can be plainly seen with an

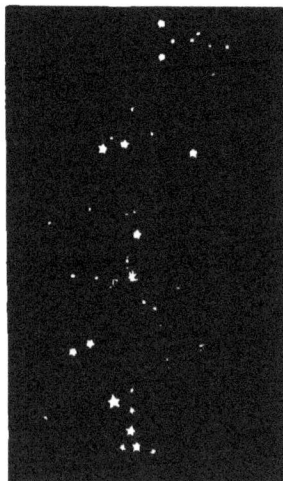

8

opera-glass. Our picture of the Sword of Orion shows its appearance with a good field-glass. With such a glass several fine test-objects will be found in the Sword. One of the best of these is formed by the two five-pointed stars seen in the picture close together above the nebula. No difficulty will be encountered in separating these stars with a field-glass, but it will require a little sharp watching to detect the small star between the two and just above the line joining them. So, the bending row of faint stars above and to the right of the group just described will be found rather elusive as individuals, though easily glimpsed as a whole. Of the great nebula itself not much detail can be seen. Yet by averting the eyes the extension of the nebulous light in every direction from the center can be detected, and traced, under favorable circumstances, to a considerable distance. The changes that this nebula certainly has undergone in the brilliancy, if not in the form, of different parts of it, are perhaps indications of the operation of forces, which we know must prevail there, and whose tendency can only be in the direction of condensation, and the ultimate formation of future suns and worlds. Yet, as the appearance of the nebula in great telescopes shows, we can not expect that the processes of creation will here produce a homologue of our solar system. The curdled appearance of the nebula indicates the formation of various centers of condensation, the final result of which will doubtless be a group of stars like some of those which we see in the heavens, and whose common motion shows that they are bound together in the chains of reciprocal gravitation. The Pleiades are an example of such a group.

Do not fail to look for a little star just west of Rigel, which, with a good opera-glass, appears to be almost hidden in the flashing rays of its brilliant companion. If you have also a field-glass, after you have detected this shy little twinkler with your opera-glass, try the larger glass upon it.

You will find then that the little star originally seen is not the only one there. A still smaller star, which had before been completely hidden, will now be perceived. I may add that, with telescopes, Rigel is one of the most beautiful double stars in the sky, having a little blue companion close under its wing. Run your glass along the line of little stars ˊ forming the lion's skin or shield that Orion opposes to the onset of Taurus. Here you will find some interesting combinations, and the star marked on the map π' will especially attract your eye, because it is accompanied, about fifteen minutes to the northwest, by a seventh-magnitude star of a rich orange hue.

Look next at the little group of three stars forming the head of Orion. Although there is no nebula here, yet these stars, as seen with the naked eye, have a remarkably nebulous look, and Ptolemy regarded the group as a nebulous star. The largest star is called Lambda (λ); the others are Phi (ϕ) one and two. An opera-glass will show another star above (λ), and a fifth star below ϕ^2, which is the farthest of the two Phis from Lambda. It will also reveal a faint twinkling between λ and ϕ'. A field-glass shows that this twinkling is produced by a pretty little row of three stars of the eighth and ninth magnitudes.

In fact, Orion is such a striking object in the sky that more than one attempt has been made to steal away its name and substitute that of some modern hero. The University of Leipsic, in 1807, formally resolved that the stars forming the Belt and Sword of Orion should henceforth be known as the constellation of Napoleon. As if to offset this, an Englishman proposed to rename Orion for the British naval bull-dog Nelson. But "Orion armed" has successfully maintained his name and place against all comers. As becomes the splendor of his constellation, Orion is a tremendous hero of antiquity, although it must be confessed that his history is somewhat shadowy and uncertain, even for a mythological

story. All accounts agree, however, that he was the mightiest hunter ever known, and the Hebrews claimed that he was
no less a person than Nimrod himself.

The little constellations of Lepus and Columba, below
Orion, need not detain us long. You will find in them some

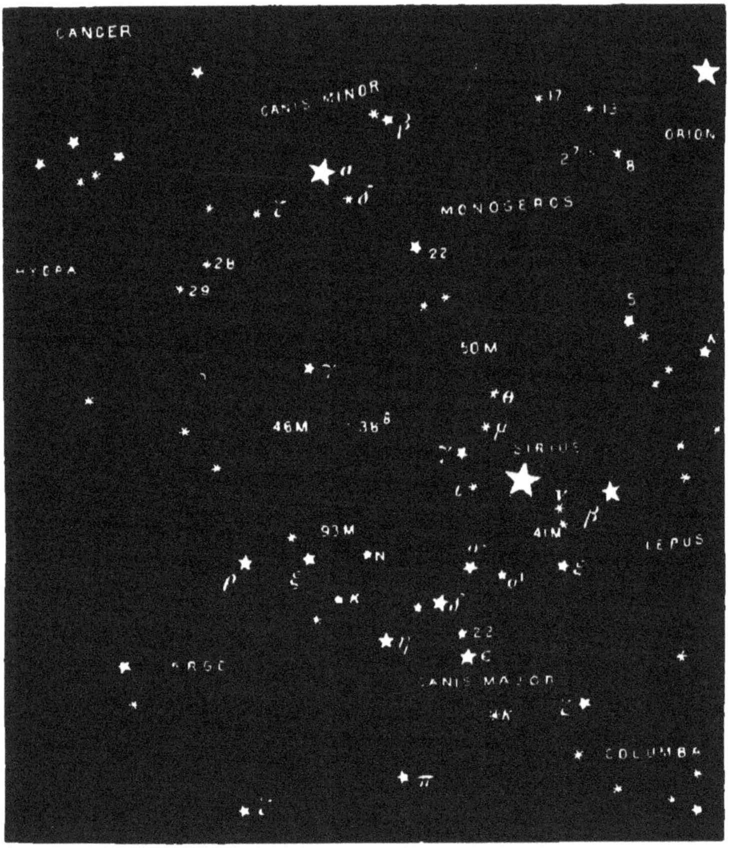

MAP 20.

pretty combinations of stars. In Lepus is the celebrated
"Crimson Star," which has been described as resembling a
drop of blood in color—a truly marvelous hue for a sun—
but, as it is never brighter than the sixth magnitude, and

from that varies down to the ninth, we could hardly hope to
see its color well with an opera-glass. Besides, the observer
would have difficulty in finding it.

We will now turn to the constellation of Canis Major, represented in Map No. 20. Although, as a constellation, it is
not to be compared with the brilliant Orion, yet, on account
of the unrivaled magnificence of its chief star, Canis Major
presents almost as attractive a scene as its more extensive
rival. Everybody has heard of Sirius, or the Dog-Star, and
everybody must have seen it flashing and scintillating so
splendidly in the winter heavens, that to call it a first-magnitude star does it injustice, since no other star of that magnitude is at all comparable with it. Sirius, in fact, stands in a
class by itself as the brightest star in the sky. Its light is
white, with a shade of green, which requires close watching to
be detected. When it is near the horizon, or when the atmosphere is very unsteady, Sirius flashes prismatic colors
like a great diamond. The question has been much dis-
cussed, as to whether Sirius was formerly a red star. It
is described as red by several ancient authors, but it seems
to be pretty well established that these descriptions are most
of them due to a blunder made by Cicero in his translation
of the astronomical poem of Aratus. It is not impossible,
though it is highly improbable, that Sirius has changed
color.

So intimately was Sirius connected in the minds of the
ancient Egyptians with the annual rising of the Nile, that it
was called the Nile-star. When it appeared in the morning
sky, just before sunrise, the season of the overflowing of the
great river was about to begin, and so the appearance of this
star was regarded as foretelling the coming of the floods.
The dog-days got their name from Sirius, as they occur at
the time when that star rises with the sun.

Your eyes will be fairly dazzled when you turn your glass
upon this splendid star. By close attention you will be able

to perceive a number of faint stars, mere points by comparison, in the immediate neighborhood of Sirius.   There are
many interesting ob-
jects in the constella-
tion. The star marked
Nu (ν) in the map is
really triple, as the
smallest glass will show.
Look next at the star-
group 41 M. The cloud
of minute stars of which
it is composed can be
very well seen with a
field-glass or a powerful
opera - glass.   The star
22 is of a very ruddy
color   that   contrasts
beautifully with the light of Epsilon (ε), which can be seen
in the same field of view with an opera-glass.   Between the
stars Delta (δ) and o' and o² there is a remarkable array of
minute stars, as shown in the accompanying cut.   One never
sees stars arranged in streams or rows, like these, without
an irresistible impression that the arrangement can not be
accidental ; that some law must have been in operation which
associated them together in the forms which we see.   Yet,
when we reflect that these are all suns, how far do we seem
to be from understanding the meaning of the universe !

The extraordinary size and brilliancy of Sirius might
naturally enough lead one to suppose that it is the nearest of
the stars, and such it was once believed to be.   Observations
of stellar parallax, however, show that this was a mistake.
The distance of Sirius is so great that no satisfactory determi-
nation of it has yet been made.   We may safely say, though,
that that distance is, at the least calculation, 50,000,000,000,-
000 miles.   In other words, Sirius is about 537,000 times as

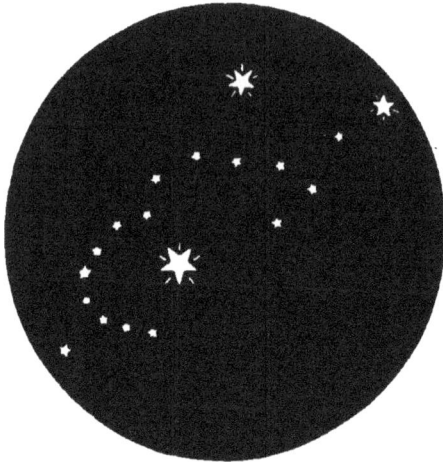

DELTA CANIS MAJORIS AND ITS NEIGHBORS.

far from the earth as the sun is. Then, since light diminishes as the square of the distance increases, the sun, if placed as far from us as Sirius is, would send us, in round numbers, 288,000,000,000 times less light than we now receive from it. But Sirius actually sends us only about 4,000,000,000 times less light than the sun does ; consequently Sirius must shine $\frac{288,000,000,000}{4,000,000.000}=72$ times as brilliantly as the sun. If we adopt Wollaston's estimate of the light of Sirius, as compared with that of the sun, viz., $\frac{1}{20.000.000,000}$, we shall still find that the actual brilliancy of that grand star is more than fourteen times as great as that of our sun. But as observations on the companion of Sirius show that Sirius's mass is fully twenty times the sun's, and since the character of Sirius's spectrum indicates that its intrinsic brightness, surface for surface, is much superior to the sun's, it is probable that our estimate of the star's actual brilliancy, as compared with what the sun would possess at the same distance, viz., seventy-two times, is much nearer the truth. It is evident that life would be insupportable upon the earth if it were placed as near to Sirius as it is to the sun. If the earth were a planet belonging to the system of Sirius, in order to enjoy the same amount of heat and light it now receives, it would have to be removed to a distance of nearly 800,000,000 miles, or eight and a half times its distance from the sun. Its time of revolution around Sirius would then be nearly five and a half years, or, in other words, the year would be lengthened five and a half times.

But, as I have said, the estimate of Sirius's distance used in these calculations is the smallest that can be accepted. Good authorities regard the distance as being not less than 100,000,000,000,000 miles ; in which case the star's brilliancy must be as much as 228 times greater than that of the sun ! And yet even Sirius is probably not the greatest sun belonging to the visible universe. There can be little doubt that

Canopus, in the southern hemisphere, is a grander sun than Sirius. To our eyes, Canopus is only about half as bright as Sirius, and it ranks as the second star in the heavens in the order of brightness. But while Sirius's distance is measurable, that of Canopus is so unthinkably immense that astronomers can get no grip upon it. If it were only twice as remote as Sirius, it would be equal to two of the latter, but in all probability its distance is much greater than that. And possibly even Canopus is not the greatest gem in the coronet of creation.

Sirius, as we saw when talking of Procyon (see Chapter I), is a double star. For many years after Bessel had declared his belief that the Dog-Star was subjected to the attraction of an invisible companion, telescopes failed to reveal the accompanying star.* Finally, in 1862, a new telescope that Alvan Clark had just finished and was testing, brought the hidden star into view. The suggestion that it may shine by reflected light from Sirius has been made. In that case it must, of course, be a planet, but a planet of such stupendous magnitude that the imagination can scarcely grasp it ; a planet probably as large as our sun, perhaps larger ; a planet equal in size to more than a million earths ! But, as was remarked of the faint stars in Alpha Capricornis, it is probable that the hypothesis of reflected light is not the true one. More probably the companion of Sirius shines with light of its own, though its excessive faintness in comparison with its bulk indicates that its condition must be very different from that of an ordinary star.

* The following extract from a letter by Bessel to Humboldt, written in 1844 (see " Cosmos," vol. iii, p. 186), is interesting, in view of the discoveries made since then : " At all events I continue in the belief that Procyon and Sirius are true double stars, consisting of a visible and an invisible star. No reason exists for considering luminosity an essential property of these bodies. The fact that numberless stars are visible is evidently no proof against the existence of an equally incalculable number of invisible ones. The physical difficulty of a change in the proper motion is satisfactorily set aside by the hypothesis of dark stars."

Readers of Voltaire will remember that the hero of his extraordinary story of "Micromegas" came from an imaginary planet circling around Sirius. Inasmuch as Voltaire, together with Dean Swift, ascribed two moons to Mars many years before they were discovered (probably suggested by a curiously mistaken interpretation by Kepler of an anagram in which Galileo had concealed his discovery of the ring of Saturn), it is all the more interesting that the great infidel should have imagined an enormous planet circling around the Dog-Star. But Voltaire went far astray when he ascribed a gigantic stature to his "Sirian." He makes Micromegas, whose world was 21,600,000 times larger in circumference than the earth, more than twenty miles tall, so that when he visited our little planet he was able to wade through the oceans and step over the mountains without inconvenience, and, when he had scooped up some of the inhabitants on his thumb-nail, was obliged to use a powerful microscope in order to see them. Voltaire should rather have gone to some of the most minute of the asteroids for his giant, for under the tremendous gravitation of such a world as he has described Micromegas himself would have been a fit subject for microscopic examination. But, however much we may doubt the stature of Voltaire's visitor from Sirius, we can not doubt the soundness of the conclusion at which he arrived, after having, by an ingenious arrangement, succeeded in holding a conversation with some earthly philosophers under his microscope, namely, that these infinitely little creatures possessed a pride that was almost infinitely great.

East and south of Canis Major, which, by-the-way, is said to represent one of Orion's hounds, is part of the constellation Argo, which stands for the ship in which Jason sailed in search of the golden fleece. The observer will find many objects of interest here, although some of them are so close to the horizon in our latitudes that much of their brilliancy is lost. Note the two stars ζ and π near the lower edge of the map, then sweep slowly over the space lying between them. About half-way your attention will be arrested by a remarkable stellar arrangement, in which a beautiful half-circle of small stars curving above a larger star, which is reddish in color, is conspicuous. This neighborhood will be found rich in stars that the naked eye can not see. Just below the star η, in Canis Major, is another fine group. The star π, which is deep yellow or orange, has three little stars above it, two of which form a pretty pair. The star ξ has a companion, which forms a fine test for an opera-glass, and is well worth looking for. Look also at the cluster 93 M, just above and to the west of ξ. The stars μ and κ are seen double with an opera-glass.

The two neighboring clusters, 46 M and 38', are very interesting objects. To see them well, use a powerful field-glass. A " fiery fifth-magnitude star," as Webb calls it, can be seen in the field at the same time. The presence of the Milky-Way is manifest by the sprinkling of stars all about this region. In fact, the attentive observer will before this have noticed that the majority of the most brilliant constellations lie either in the Milky-Way or along its borders. Cassiopeia, as we saw, sits athwart the galaxy whose silvery current winds in and out among the stars of her "chair"; Perseus is aglow with its sheen as it wraps him about like a mantle of stars; Taurus has the tips of his horns dipped in the great stream; it flows between the shining feet of Gemini and the head and shoulders of Orion as between starry banks; the peerless Sirius hangs like a gem pendent from the celestial girdle. In the southern hemisphere we should find the beautiful constellation of the ship Argo, containing Canopus, sailing along the Milky-Way, blown by the breath of old romance on an endless voyage; the Southern Cross glitters in the very center of the galaxy; and the bright stars of the Centaur might be likened to the heads of golden nails pinning this wondrous scarf, woven of the beams of millions of tiny stars, against the dome of the sky. Passing back into the northern hemisphere we find Scorpio, Sagittarius, Aquila, the Dolphin, Cygnus, and resplendent Lyra, all strung along the course of the Milky-Way.

Turning now to the constellation Monoceros, we shall find a few objects worthy of attention. This constellation is of comparatively modern origin, having been formed by Bartschius, whose chief title to distinction is that he married the daughter of John Kepler. The region around the stars 8, 13, and 17 will be found particularly rich, and the cluster 2' shows well with a strong glass. Look also at the cluster 50 M, and compare its appearance with that of the clusters in Argo.

With these constellations we finish our review of the stellar wonders that lie within the reach of so humble an instrument as an opera- or field-glass. We have made the circuit of the sky, and the hosts that illumine the vernal heavens are now seen advancing from the east, and pressing close upon the brighter squadrons of winter. Their familiar figures resemble the faces of old friends whom we are glad to welcome. These starry acquaintances never grow wearisome. Their interest for us is as fathomless as the deeps of space in which they shine. The man never yet lived whose mind could comprehend the full meaning of the wondrous messages that they flash to us upon the wings of light. As we watch them in their courses, the true music of the spheres comes to our listening ears, the chorus of creation—faint with distance, for it is by slow approaches that man draws near to it—chanting the grandest of epics, the Poem of the Universe ; and the theme that runs through it all is the reign of law. Do not be afraid to become a star-gazer. The human mind can find no higher exercise. He who studies the stars will discover—

> " An endless fountain of immortal drink
> Pouring unto us from heaven's brink."

CHAPTER V

"IT is a most beautiful and delightful sight," exclaims Galileo, in describing the discoveries he had made with his telescope, "to behold the body of the moon, which is distant from us nearly sixty semi-diameters of the earth, as near as if it was at a distance of only two of the same measures. . . . And, consequently, any one may know with the certainty that is due to the use of our senses that the moon assuredly does not possess a smooth and polished surface, but one rough and uneven, and, just like the face of the earth itself, is everywhere full of vast protuberances, deep chasms, and sinuosities."

There was, perhaps, nothing in the long series of discoveries with which Galileo astonished the world after he had constructed his telescope, which, as he expresses it, "was devised by me through God's grace first enlightening my mind," that had a greater charm for him than his lunar observations. Certainly there was nothing which he has described with greater enthusiasm and eloquence. And this could hardly have been otherwise, for the moon was the first celestial object to which Galileo turned his telescope, and then for the first time human eyes may be said to have actually looked into another world than the earth, though his discoveries and those of his successors have not realized all the poetic fancies about the moon contained in the verses that are ascribed to Orpheus:

"And he another wandering world has made
Which gods Selene name, and men the moon.
It mountains, cities has, and temples grand."

Yet Galileo's observations at once upset the theory, for which Apollonius was responsible, and which seems to have been widely prevalent up to his time, that the moon was a smooth body, polished like a mirror, and presenting in its light and dark spots reflections of the continents and oceans of the earth. He also demonstrated that its surface was covered with plains and mountains, but the "cities and temples" of the moon have remained to our time only within the ken of romance.

Galileo's telescope, as I have before remarked, was, in the principle of its construction, simply an opera-glass of one tube. He succeeded in making a glass of this kind that magnified thirty diameters, a very much higher power than is given to the opera- and field-glasses of to-day. Yet he had to contend with the disadvantages of single lenses, achromatic combinations of glass for optical purposes not being contrived until nearly a hundred years after his death, and so his telescope did not possess quite as decided a superiority over a modern field-glass as the difference in magnifying power would imply. In fact, if the reader will view the moon with a first-rate field-glass, he will perceive that the true nature of the surface of the lunar globe can be readily discerned with such an instrument. Even a small opera-glass will reveal much to the attentive observer of the moon; but for these observations the reader should, if possible, make use of a field-glass, and the higher its power the better. The illustrations accompanying this chapter were made by the author with the aid of a glass magnifying seven diameters.

Of course, the first thing the observer will wish to see will be the mountains of the moon, for everybody has heard of them, and the most sluggish imagination is stirred by the thought that one can look off into the sky and behold "the

eternal hills " of another planet as solid and substantial as
our own. But the chances are that, if left to their own
guidance, ninety-nine persons out of a hundred would choose
exactly the wrong time to see these mountains. At any rate,
that is my experience with people who have come to look at
the moon through my telescope. Unless warned beforehand,
they invariably wait until full moon, when the flood of sunshine poured perpendicularly upon the face of our satellite
conceals its rugged features as effectually as if a veil had
been drawn over them. Begin your observations with the
appearance of the narrowest crescent of the new moon, and
follow it as it gradually fills, and then you will see how beautifully the advancing line of lunar sunrise reveals the mountains, over whose slopes and peaks it is climbing, by its ragged
and sinuous outline. The observer must keep in mind the
fact that he is looking straight down upon the tops of the
lunar mountains. It is like a view from a balloon, only at a
vastly greater height than any balloon has ever attained.
Even with a powerful telescope the observer sees the moon
at an apparent distance of several hundred miles, while with
a field-glass, magnifying seven diameters, the moon appears as
if thirty-five thousand miles off. The apparent distance with
Galileo's telescope was eight thousand miles. Recollect how
when seen from a great height the rugosities of the earth's
surface flatten out and disappear, and then try to imagine
how the highest mountains on the earth would look if you
were suspended thirty-five thousand miles above them, and
you will, perhaps, rather wonder at the fact that the moon's
mountains can be seen at all.

It is the contrast of lights and shadows that not only reveals them to us, but enables us to measure their height.
On the moon shadows are very much darker than upon the
earth, because of the extreme rarity of the moon's atmosphere, if indeed it has any atmosphere at all. By stepping
around the corner of a rock there, one might pass abruptly

from dazzling noonday into the blackness of midnight. The surface of the moon is extraordinarily rough and uneven. It possesses broad plains, which are probably the bottoms of ancient seas that have now dried up, but these cover only about two fifths of the surface visible to us, and most of the remaining three fifths are exceedingly rugged and mountainous. Many of the mountains of the moon are, foot for foot, as lofty as the highest mountains on the earth, while all of them, in proportion to the size of the moon's globe, are much larger than the earth's mountains. It is obvious, then, that the sunshine, as it creeps over these Alpine landscapes in the moon, casting the black shadows of the peaks and craters many miles across the plains, and capping the summits of lofty mountains with light, while the lower regions far around them are yet buried in night, must clearly reveal the character of the lunar surface. Mountains that can not be seen at all when the light falls perpendicularly upon them, or, at the most, appear then merely as shining points, picture themselves by their shadows in startling silhouettes when illuminated laterally by the rising sun.

But at full moon, while the mountains hide themselves in light, the old sea-beds are seen spread out among the shining table-lands with great distinctness. Even the naked eye readily detects these as ill-defined, dark patches upon the face of the moon, and to their presence are due the popular notions that have prevailed in all quarters of the world about the "Man in the Moon," the "Woman in the Moon," "Jacob in the Moon," the "Hare in the Moon," the "Toad in the Moon," and so on. But, however clearly one may imagine that he discerns a man in the moon while recalling the nursery-rhymes about him, an opera-glass instantly puts the specter to flight, and shows the round lunar disk diversified and shaded like a map.*

* I should, perhaps, qualify the statement in the text slightly in favor of a lunar lady to whom Mr. Henry M. Parkhurst first called my attention. About

A feature of the full moon's surface that instantly attracts attention is the remarkable brightness of the southern part of the disk, and the brilliant streaks radiating from a bright point near the lower edge. The same simile almost invariably comes to the lips of every person who sees this phenomenon for the first time—"It looks like a peeled orange." The bright point, which is the great crater-mountain Tycho, looks exactly like the pip of the orange, and the light-streaks radiating from it in all directions bear an equally striking resemblance to the streaks that one sees upon an orange after the outer rind has been removed. I shall have something more to say about these curious streaks further on ; in the mean time, let us glance at our little sketch-map of the moon.

The so-called seas are marked on the map, for the purpose of reference, by the letters which they ordinarily bear in lunar maps. The numerals indicate craters, or ring-plains, and mountain-ranges. The following key-list will enable the reader to identify all the objects that are lettered or numbered upon the map. I have given English translations of the Latin names which the old astronomers bestowed upon the seas :

nine days after new moon a rather pretty and decidedly feminine face appears on the western half of the disk. It is formed by the mountains and table-lands embraced by the Sea of Serenity, the Sea of Tranquillity, the Sea of Vapors, etc., and is best seen with the aid of an opera-glass of low power. The face is readily distinguishable on Rutherfurd's celebrated photograph of the full moon. It is necessary for this purpose to turn the photograph upside down, since it is a telescopic picture, and consequently reversed. The crater Tycho forms a breastpin for the lady, and Menelaus glitters like a diamond ornament in her hair, while the range of the Apennines resembles a sort of coronet resting on her forehead. This same woman in the moon, it appears, was described by Dr. James Thompson years ago, and, for aught I know, she may be the Diana to whom Herrick sang:

"Queen and huntress chaste and fair,
    Seated in thy silver chair,
    Now the Sun is laid to sleep,
    State in wonted manner keep.
    Hesperus entreats thy light,
    Goddess excellently bright."

NORTH.

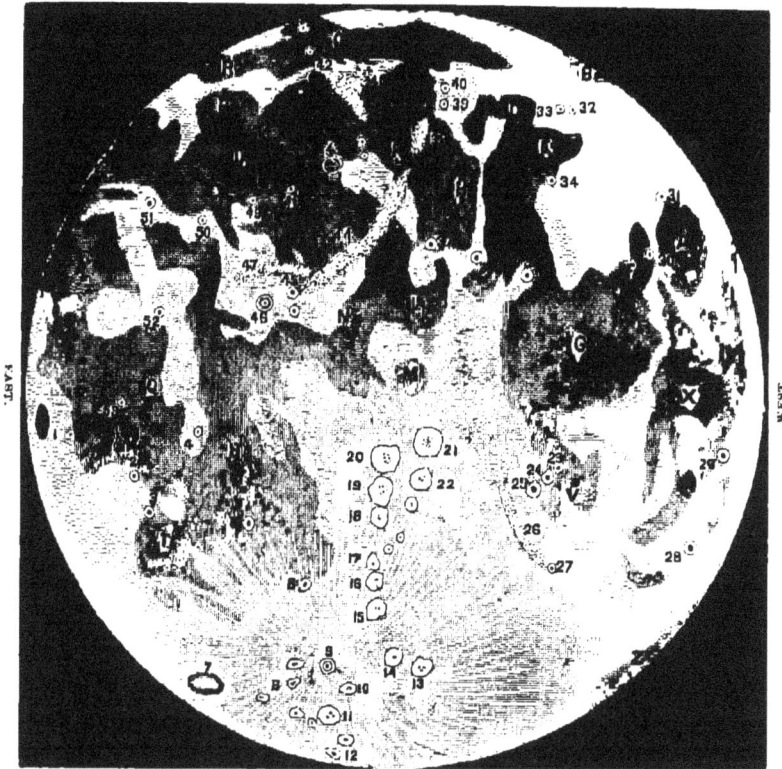

SOUTH.

MAP OF THE MOON.

Seas, Gulfs, and Marshes.

A. The Crisian Sea.	I. The Marsh of Mists.	R. The Bay of Dew.
B. Humboldt Sea.	K. The Marsh of Putrefaction.	S. The Sea of Clouds.
C. The Sea of Cold.	L. The Sea of Vapors.	T. The Sea of Humors.
D. The Lake of Death.	M. The Central Gulf.	V. The Sea of Nectar.
E. The Lake of Dreams.	N. The Gulf of Heats.	X. The Sea of Fertility.
F. The Marsh of Sleep.	O. The Sea of Showers.	Z. The South Sea.
G. The Sea of Tranquillity.	P. The Bay of Rainbows.	
H. The Sea of Serenity.	Q. The Ocean of Storms.	

Mountains and Crater Rings.

1. Grimaldi.	15. Walter.	28. Petavius.	41. The Alps.
2. Letronne.	16. Regiomontanus.	29. Langrenus.	42. Plato.
3. Gassendi.	17. Purbach.	30. Proclus.	43. Archimedes.
4. Euclides.	18. Arzachel.	31. Cleomedes.	44. The Apennines.
5. Bullialdus.	19. Alphonsus.	32. Atlas.	45. Eratosthenes.
6. Pitatus.	20. Ptolemaus.	33. Hercules.	46. Copernicus.
7. Schickhard.	21. Hipparchus.	34. Posidonius.	47. The Carpathian Mts.
8. Longomontanus.	22. Albategnius.	35. Plinius.	48. Timocharis.
9. Tycho.	23. Theophilus.	36. Menelaus.	49. Lambert.
10. Maginus.	24. Cyrillus.	37. Manilius.	50. Euler.
11. Clavius.	25. Catharina.	38. The Caucasus Mts.	51. Aristarchus.
12. Newton.	26. The Altai Mts.	39. Eudoxus.	52. Kepler.
13. Maurolycus.	27. Piccolomini.	40. Aristotle.	53. Flamsteed.
14. Stöfler.			

9

The early selenographers certainly must have been men of vivid imagination, and the romantic names they gave to the lunar landscapes, and particularly to the "seas," add a charm of their own to the study of the moon. Who would not wish to see the "Bay of Rainbows," or the "Lake of Dreams," or the "Sea of Tranquillity," if for no other reason than a curiosity to know what could have induced men to give to these regions in the moon such captivating titles? Or who would not desire to visit them if he could? though no doubt we should find them, like the "Delectable Mountains" in the "Pilgrim's Progress," most charming when seen from afar.

The limited scale of our map, of course, renders it impossible to represent upon it more than a comparatively small number of the lunar mountains that have received names. In selecting those to be put in the map I have endeavored to choose such as, on account of their size, their situation, or some striking peculiarity, would be most likely to attract the attention of a novice. The observer must not expect to see them all at once, however. The lunar features change their appearance to a surprising extent, in accordance with the direction of their illumination. Some great mountain-masses and ring-plains, or craters, which present scenes of magnificence when the sun is rising or setting upon them, disappear under a perpendicular light, such as they receive at full moon. The great crater-plain, known as Maginus, numbered 10 in our map, is one of these. The broken mountain-wall surrounding this vast depressed plain rises in some places to a height of over fourteen thousand feet above the valley within, and the spectacle of sunrise upon Maginus, seen with a powerful telescope, is a most impressive sight, and even with a field-glass is very interesting. Yet, a few days later, Maginus vanishes, as if it had been swallowed up, and as Beer and Mädler have expressed it, "the full moon knows no Maginus." The still grander formation of mountain, plain, and crater, called Clavius (11 in the map), disappears almost as completely as Maginus at full moon, yet, under the proper illumination, it presents a splendid pageant of light and shadow.

On the other hand, some of the lunar mountains shine vividly at full moon, and can be well seen then, though, of course, only as light spots, since at that time they cast no shadows. Menelaus (36 in the map), Aristarchus (51), Proclus (30), Copernicus (46), and Kepler (52), are among these shining mountains. Aristarchus is the most celebrated of them all, being the brightest point on the moon. It can even be seen glimmering on the dark side of the moon—that is to say, when no light reaches it except that which is reflected from the earth. With a large telescope, Aristarchus is so dazzlingly bright under a high sun, that the eye is partly blinded in gazing at it. It consists of a mountain-ring surrounding a circular valley, about twenty-eight miles in diameter. The flanks of these mountains, especially on their inner slopes, and the floor of the valley within, are very bright, while a peak in the center of the valley, about as high as Storm-King Mountain on the Hudson, shines with piercing brilliancy. Sir William Herschel mistook it for a volcano in action. It certainly is not an active volcano, but just what makes it so dazzling no one knows. The material of which this mountain is formed would seem to possess a higher reflective power than that of any other portion of the moon's surface. One is irresistibly reminded of the crystallized mountains described in the celebrated "Moon Hoax" of Richard Adams Locke. With an opera-glass you can readily recognize Aristarchus as a bright point at full moon. With a field-glass it is better seen, and some of the short, light rays surrounding it are perceived, while, when the sun is rising upon it, about four days after first quarter, its crateriform shape can be detected with such a glass.

The visibility of Aristarchus on the dark side of the moon leads us to a brief consideration of the illumination by the earth of that portion of the moon's surface which is not touched directly by sunlight at new and old moon. This phenomenon is shown in the accompanying illustration. Not only can the outlines of the dark part of the moon be seen under such circumstances, but even the distinction in color between the dusky "seas" and the more brilliant table - lands and mountain - regions can be perceived, and with powerful

SUNRISE ON THE SEA OF SERENITY, AND THEOPHILUS AND OTHER CRATERS.

telescopes many minor features come into sight. A little consideration must convince any one, as it convinced Galileo more than two hundred and seventy-five years ago, that the light reflected from the earth upon the moon is sufficient to produce this faint illumination of the lunar landscapes. We have only to recall the splendors of a night that is lighted by a full moon, and then to recollect that at new or old moon the earth is "full" as seen from our satellite, and that a full earth must give some fourteen times as much light as a full moon, in order to realize the brilliancy of an earth-lit night upon the moon. As the moon waxes to us, the earth wanes to the moon, and *vice versa*, and so the phenomenon of earth-shine on the lunar surface must be looked for before the first quarter and after the last quarter of the moon.

    The reader will find it an attractive occupation to identify, by means of the map, the various "seas," "lakes," and "marshes," for not only are they interesting on account of the singularity of their names, but they present many remarkable differences of appearance, which may be perceived with the instrument he is supposed to be using. The oval form of the Crisian Sea (A), which is the first of the "seas" to come into sight at new moon, makes it a very striking object. With good telescopes, and under favorable illumination, a decidedly green tint is perceived in the Crisian Sea. It measures about two hundred and eighty by three hundred and fifty-five miles in extent, and is, perhaps, the deepest of all the old sea-beds visible on the moon. It is surrounded by mountains, which can be readily seen when the sun strikes athwart them a few days after new or full moon. On the southwestern border a stupendous mountain-promontory, called Cape Agarum, projects into the Crisian Sea fifty or sixty miles, the highest part rising precipitously eleven thousand feet above the floor of the sea. I have seen Cape Agarum very clearly defined with a field-glass. Near the eastern border is the crater-mountain Proclus, which I have already mentioned as possessing great brilliancy under a high sun, being in this respect second only to Aristarchus.

From the foot of Proclus spreads away the somewhat triangular region called the Marsh of Sleep (F). The term "golden-brown," which has been applied to it, perhaps describes its hue well enough. With a telescope it is a most interesting region, but with less powerful instruments one must be content with recognizing its outline and color.

The broad, dark-gray expanse of the Sea of Tranquillity (G) will be readily recognized by the observer, and he will be interested in the mottled aspect which it presents in certain regions, caused by ridges and elevations, which, when this sea-bottom was covered with water, may have formed shoals and islands.

The Sea of Fertility (X) is remarkable for its irregular

surface, and the long, crooked bays into which its southern extremity is divided.

The Sea of Nectar (V) is connected with the Sea of Tranquillity by a broad strait (one would naturally anticipate from their names that there must be some connection between them), while between it and the Sea of Fertility runs the range of the Pyrenees Mountains, twelve thousand feet high, flanked by many huge volcanic mountain-rings.

The Sea of Serenity (H), lying northeast of the Sea of Tranquillity, is about four hundred and twenty miles broad by four hundred and thirty miles long, being very nearly of the same area as our Caspian Sea. It is deeper than the Sea of Tranquillity, and a greenish hue is sometimes detected in its central parts. It deepens toward the middle. Three quarters of its shore-line are bordered by high mountains, and many isolated elevations and peaks are scattered over its surface. In looking at these dried-up seas of the moon, one is forcibly reminded of the undulating and in some places mountainous character of terrestrial sea-bottoms, as shown by soundings and the existence of small islands in the deep sea, like the Bermudas, the Azores and St. Helena. The Sea of Serenity is divided nearly through the center by a narrow, bright streak, apparently starting from the crater-mountain Menelaus (36 in the map), but really taking its rise at Tycho far in the south. This curious streak can be readily detected even with a small opera-glass. Just what it is no one is prepared to say, and so the author of the "Moon Hoax" was fairly entitled to take advantage of the romancer's license, and declare that "its edge throughout its whole length of three hundred and forty miles is an acute angle of solid quartz-crystal, brilliant as a piece of Derbyshire spar just brought from the mine, and containing scarcely a fracture or a chasm from end to end!" Along the southern shore, on either side of Menelaus, extends the high range of the Hæmus Mountains. South and southeast of the Sea of Serenity are the Sea of Vapors (L), the Central Gulf (M), and the Gulf of Heats (N). The observer will notice at full moon three or four curious dark spots in the region occupied by these flat expanses. On the north and northwest of the Sea of Serenity are the Lake of Death (D), and the Lake of Dreams (E), chiefly remarkable for their names.

The Sea of Showers (O) is a very interesting region, not only in itself, but on account of its surroundings. Its level is very much broken by low, winding ridges, and it is variegated by numerous light-streaks. At its western end it blends into the Marsh of Mists (I) and the Marsh of Putrefaction (K). On its northeast border is the celebrated Sinus Iridum, or Bay of Rainbows (P), upon which selenographers have exhausted the adjectives of admiration. The bay is semicircular in form, one hundred and thirty-five miles long and eighty-four miles broad. Its surface is dark and level. At either end a splendid cape extends into the Sea of Showers, the eastern one being called Cape Heraclides, and the western Cape Laplace. They are both crowned by high peaks. Along the whole shore of the bay runs a chain of gigantic mountains, forming the southern border of a wild and lofty plateau, called the Sinus Iridum Highlands. Of course, a telescope is required to see the details of this "most magnificent of all lunar landscapes," and yet much can be done with a good field-glass. With such an instrument I have seen the capes at the ends of the bay projecting boldly into the dark, level expanse surrounding them, and the high lights of the bordering mountains sharply contrasted with the dusky semicircle at their feet, and have been able to detect the presence of the low ridges that cross the front of the bay like shoals, separating it from the "sea" outside. Two or three days after first quarter, the shadows of the peaks about the Bay of Rainbows may be seen. The Bay of Dew (R) above the Bay of Rainbows, and the Sea of Cold (C), are the northernmost of the dark levels visible. It was in keeping with the supposed character of this region of the Moon that Riccioli named two portions of it the Land of Hoar Frost and the Land of Drought.

Extending along the eastern side of the disk is the great Ocean of Storms (Q), while between the Ocean of Storms and the middle of the moon lies the Sea of Clouds (S). Both of these are very irregular in outline, and much broken by ridges and mountains. The Sea of Humors (T), although comparatively small, is one of the most easily seen of all the lunar plains. To the naked eye it looks like a dark, oval patch on the moon. With a telescope it is seen, under favorable conditions, to possess a decided green tint. Humboldt Sea (B) and the South Sea (Z) belong principally to that part of the moon which is always turned away from the earth, and only their edges project into the visible hemisphere, although, under favorable librations, their farther borders, lined as usual with mountain-peaks, may be detected. For our purposes they possess little interest.

Let us now glance at some of the mountains and "craters." The dark oval called Grimaldi (1) can be detected by the naked eye, or at least it has been thus seen, although it requires a sharp eye; and perhaps a shade or a pair of eye-glasses of London smoke-glass, to take off the glare of the moon, should be used in looking for it.* It is simply a plain, containing some fourteen thousand square miles, remarkable for its dark color, and surrounded by mountains. Schickhard (7) is another similar plain, nearly as large, but not possessing the same dark tint in the interior. The huge mountains around Schickhard make a fine spectacle when the sun is rising upon them shortly before full moon.

* There are other uses to which such eye-glasses may be put by sky-gazers. I habitually carry a pair for studying clouds. It is wonderful how much the effect of great cloud-masses is heightened by them, especially when seen in a bright light. Delicate curls and striæ of cirrus, which escape the uncovered eye in the glare of sunlight, can be readily detected and studied by the use of neutral-tinted eye-glasses or spectacles.

Tycho (9) is the most famous of the crater-mountains, though not the largest. It is about fifty-four miles across and three miles deep. In its center is a peak five or six thousand feet high. Tycho is the radial point of the great light-streaks that, as I have already remarked, cause the southern half of the moon to be likened to a peeled orange. It is a tough problem in selenography to account for these streaks. They are best seen at full moon. They can not be seen at all until the sun has risen to a certain elevation above them, 25° according to Neison; but, when they once become visible, they dominate everything. They turn aside for neither mountains nor plains, but pass straight on their courses over the ruggedest regions of the moon, retaining their brilliancy undiminished, and pouring back such a flood of reflected light that they completely conceal some of the most stupendous mountain-masses across which they lie. They clearly consist of different material from that of which the most of the moon's surface is composed—a material possessing a higher reflective power. In this respect they resemble Aristarchus and other lunar craters that are remarkable for their brilliancy under a high illumination. Tycho itself, the center or hub, from which these streaks radiate like spokes, is very brilliant in the full moon. But immediately around Tycho there is a dark rim some twenty-five miles broad. Beyond this rim the surface becomes bright, and the bright region extends about ninety miles farther. Out of it spring the great rays or streaks, which vary from ten to twenty miles in width, and many of which are several hundred miles long—one, which we have already mentioned as extending across the Sea of Serenity, being upward of two thousand miles in length. It has been truly said that we have nothing like these streaks upon the earth, and so there is no analogy to go by in trying to determine their nature. It has been suggested that if the moon had been split or shattered from within by some tremendous force, and molten matter from the interior had been thrust up into the cracks thus formed, and had cooled there into broad seams of rock, possessing a higher reflective power than the surrounding surface of the moon, then the appearances presented would not be unlike what we actually see. But there are serious objections to such a view, which we have not space to discuss here. It is enough to say that the nature of these streaks is still a question awaiting solution, and here is an opportunity for an important discovery, but not one to be achieved with an opera-glass.

I may add an interesting suggestion as to the nature of these streaks made by the Rev. Mr. Grensted. He holds that the air and water of the moon were chemically, and not mechanically, absorbed in the process of oxidation which went on at the time when her·surface temperature was above a red heat. Having a much larger surface in proportion to her bulk than the earth, the oxidation of the moon has, he thinks, extended much deeper than that of the earth, and her atmosphere and oceans have been exhausted in the process. Both the earth and the moon, he maintains, have metallic nuclei, and the streaks about Tycho and Copernicus, and some other lunar craters, may be dikes of pure and shining metal, which have escaped oxidation owing to the comparatively small supply of lunar oxygen. Upon this theory Aristarchus must be a metallic mountain.

Clavius (11) is one of the most impressive of all the lunar formations. There probably does not exist anywhere upon the earth so wild a scene upon a corresponding scale of grandeur. Of course, its details are far beyond the reach of the instrument we are supposed to be using, and yet, even with a field-glass, or a powerful opera-glass, some of its main features are visible. It is represented in our picture of the half-moon, being the lowest and largest of the ring-like forms seen at the inner edge of the illuminated half of the disk; the rays of the rising sun touching the summits of some of the peaks

in its interior have brought them into sight as a point of
light, and at the same time, reaching across the gulf within,
have lighted up the higher
slopes of the great mountain-
wall on the farther or eastern
side of the crater - valley,
making it resemble a semicircle of light projecting into
the blackness of the still un-
illuminated plains around it.
I should advise every reader
to take advantage of any opportunity that may be presented to him to see Clavius
with a powerful telescope
when the sun is either rising
or setting upon it.   Neison
has given a spirited description of the scene, as follows:

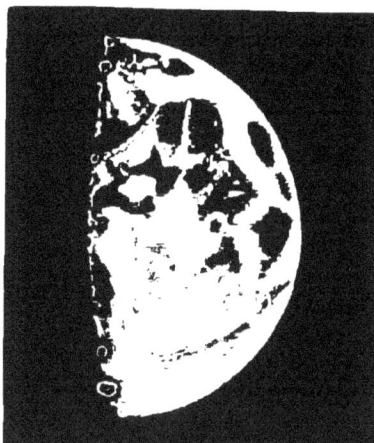

SUNRISE ON CLAVIUS, TYCHO, PLATO, ETC.

The sunrise on Clavius commences with the illumination of a few
peaks on the western wall, but soon rapidly extends along the whole
wall of Clavius, which then presents the appearance of a great double
bay of the dark night-side of the moon penetrating so deep into the
illuminated portion as to perceptibly blunt the southern horn to the
naked eye.   Within the dark bay some small, bright points soon appear—the summits of the great ring-plains within—followed shortly by
similar light-points near the center, due to peaks on the walls of the
smaller ring-plains, these light-islands gradually widening and forming delicate rings of light in the dark mass of shadow still enveloping
the floor of Clavius.   Far in the east then dimly appear a few scarcely
perceptible points, rapidly widening into a thin bright line, the crest
of the great southeastern wall of Clavius, the end being still lost far
within the night-side of the moon.   By the period the extreme sum-
mit of the lofty wall of Clavius on the east becomes distinct, fine
streaks of light begin to extend across the dark mass of shadow on the
interior of Clavius, from the light breaking through some of the
passes on the west wall and illuminating the interior; and these
streaks widen near the center and form illuminated spots on the
floor, when both east and west it still lies deeply immersed in shadow,

strongly contrasting with the now brightly illuminated crest of the lofty east wall and the great circular broad rings of light formed by the small ring-plains within Clavius. The illumination of the interior of Clavius now proceeds rapidly, and forms a magnificent spectacle : the great, brightly illuminated ring-plains on the interior, with their floors still totally immersed in shadow ; the immense steep line of cliffs on the east and southeast are now brilliantly illuminated, though the entire surface at their base is still immersed in the shades of night ; and the great peaks on the west towering above the floor are thrown strongly into relief against the dark shadow beyond them.

Newton (12) is the deepest of the great crateriform chasms on the moon. Some of the peaks on its walls rise twenty-four thousand feet above the interior gulf. Its shadow, and those of its gigantic neighbors—for the moon is here crowded with colossal walls, peaks, and craters—may be seen breaking the line of sunlight below Clavius, in our illustration. I have just spoken of these great lunar formations as chasms. The word describes very well the appearance which some of them present when the line separating day and night on the moon falls across them, but the reader should not be led by it into an erroneous idea of their real character. Such formations as Newton, which is one hundred and forty miles long by seventy broad, may more accurately be described as vast depressed plains, generally containing peaks and craters, which are surrounded by a ring of steep mountains, or mountain-walls, that rise by successive ridges and terraces to a stupendous height.

The double chain of great crater-plains reaching half across the center of the moon contains some of the grandest of these strange configurations of conjoined mountain, plain, and crater. The names of the principal ones can be learned from the map, and the reader will find it very interesting to watch them coming into sight about first quarter, and passing out of sight about third quarter. At such times, with a field-glass, some of them look like enormous round holes in the inner edge of the illuminated half of the moon. Theophilus

(23), Cyrillus (24), and Catharina (25), are three of the finest walled plains on the moon—Theophilus, in particular, being a splendid specimen of such formations. This chain of craters may be seen rapidly coming into sunlight at the edge of the Sea of Nectar, in our picture of " Sunrise on the Sea of Serenity," etc. The Altai Mountains (26) are a line of lofty cliffs, two hundred and eighty miles in length, sur-mounting a high table-land.

The Caucasus Mountains (38) are a mass of highlands and peaks, which introduce us to a series of formations resem-bling those of the mountainous regions of the earth. The highest peak in this range is about nineteen thousand feet. Between the Caucasus and the Apennines (44) lies a level pass, or strait, connecting the Sea of Serenity with the Sea of Showers. The Apennines are the greatest of the lunar mountain-chains, extending some four hundred and sixty miles in length, and containing one peak twenty-one thou-sand feet high, and many varying from twelve thousand to nearly twenty thousand. It will thus be seen that the Apen-nines of the earth sink into insignificance in comparison with their gigantic namesakes on the moon. As this range runs at a considerable angle to the line of sunrise, its high peaks are seen tipped with sunlight for a long distance beyond the generally illuminated edge about the time of first quarter. Even with the naked eye the sun-touched summits of the lunar Apennines may at that time be detected as a tongue of light projecting into the dark side of the moon. The Alps (41) are another mountain-mass of great elevation, whose high-est peak is a good match for the Mont Blanc of the earth, after which it has been named.

Plato (42) is a very celebrated dark and level plain, sur-rounded by a mountain-ring, and presenting in its inte-rior many puzzling and apparently changeable phenomena which have given rise to much speculation, but which, of course, lie far beyond the reach of opera-glasses. Plato is seen in the picture of "Sunrise on Clavius," etc., on page 133, being the second ring from the top.

If Ariosto had had a telescope, we might have suspected that it was this curious plain that he had in mind when he described that strange valley in the moon, in which was to be found everything that was lost from the earth, including lost wits; and where the redoubtable knight Astolpho, having been sent in search of the missing wit of the great Orlando, was astonished to find what he sought carefully preserved in a vial along with other similar vials belonging to many supposedly wise people of the earth, whom nobody suspected of keeping a good part of their sapience in the moon.

Copernicus (46) is the last of the lunar formations that we shall describe. It bears a general resemblance to Tycho, and is slightly greater in diameter; it is, however, not quite so deep. It has a cluster of peaks in the center, whose tops may be detected with a field-glass, as a speck of light when the rays of the morning sun, slanting across the valley, illuminate them while their environs are yet buried in night. Copernicus is the center of a system of light-streaks somewhat resembling those of Tycho, but very much shorter.

We must not dismiss the moon without a few words as to its probable condition. It was but natural, after men had seen the surface of the moon diversified with hills and valleys like another earth, that the opinion should find ready acceptance that beings not unlike ourselves might dwell upon it. Nothing could possibly have been more interesting than the realization of such a fancy by the actual discovery of the lunar inhabitants, or at least of unmistakable evidence of their existence. The moon is so near to the earth, as astronomical distances go, and the earth and the moon are so intimately connected in the companionship of their yearly journey around the sun, and their greater journey together with the sun and all his family, through the realms of space, that we should have looked upon the lunar inhabitants, if any had existed, as our neighbors over the way—dwelling, to be sure, upon a somewhat more restricted domain than ours, vassals of the earth in one sense, yet upon the whole very respectable and interesting people, with whom one would be glad to have a closer acquaintance. But, alas! as the powers of the telescope increased, the vision of a moon crowded with life faded, until at last the cold fact struck home that the moon is, in all probability, a frozen and dried-up globe, a mere planetary skeleton, which could no more support life than the Humboldt glacier could grow roses. And yet this opinion may go too far. There is reason for thinking that the moon is not absolutely airless, and, while it has no visible bodies of water, its soil may, after all, not be entirely arid and desiccated. There are observations which hint at visible changes in certain spots that could possibly be caused by vegetation, and there are other observations which suggest the display of electric luminosity in a rarefied atmosphere covering the moon. To declare that no possible form of life can exist under the conditions prevailing upon the lunar surface would be saying too much, for human intelligence can not set bounds to creative power. Yet, within the limits of life, such as we know them, it is probably safe to assert that the moon is a dead and deserted world. In other words, if a race of beings resembling ourselves, or resembling any of our contemporaries in terrestrial life, ever existed upon the moon, they must long since have perished. That such beings may have existed, is possible, particularly if it be true, as generally believed, that the moon once had a comparatively dense atmosphere and water upon its surface, which have now, in the process of cooling of the lunar globe, been withdrawn into its interior. It certainly does not detract from the interest with which we study the rugged and beautiful scenery of the moon to reflect that if we could visit those ancient sea-bottoms, or explore those glittering mountains, we might, perchance, find there some remains or mementos of a race that flourished, and perhaps was all gathered again to its fathers, before man appeared upon the earth.

That slight physical changes, such as the downfall of mountain-walls or crater-cones, still occasionally occur upon the moon, is an opinion entertained by some selenographers, and apparently justified by observation. The enormous changes of temperature, from burning heat under a cloudless sun to the freezing cold of space at night with no atmospheric blanket to retain heat (which has generally been assumed to be the condition of things on the moon), would naturally exert a disintegrating effect upon the lunar rocks. But the question is now in dispute whether the surface of the moon ever rises above the freezing-point of water, even under a midday sun.

Mankind has always been a little piqued by the impossibility of seeing the other side of the moon, and all sorts of odd fancies have been indulged in regard to it. Among the most curious is the ancient belief that the souls of the good who die on earth are transported to that side of the moon which is turned away from the earth; while the souls of the wicked sojourn on this side, in full view of the scene of their evil deeds. The visible side of the moon—with its tremendous craters, its yawning chasms, its frightful contrasts of burning sunshine and Cimmerian darkness, its airless and arid plains and dried-up sea-bottoms exposed to the pitiless cold of open space, and heated, if heated at all, by scorching sunbeams as fierce as naked flame—would certainly appear to be in a proper condition to serve as a purgatory. But we have no reason to think that the other side is any better off in these respects. In fact, the glimpses that we get of it around the corners, so to speak, indicate that the whole round globe of the moon is as ragged, barren, and terrible as that portion of it which is turned to our view.

THE PLANETS.—In attempting to view the planets with an

opera-glass, too much must not be expected ; and yet interesting views can sometimes be obtained.   The features of their
surfaces, of course, can not be detected even with a powerful
field-glass, but the difference between the appearance of a
large planet and that of the stars will at once strike the observer.   Mercury, which, on account of its nearness to the
sun and its rapid changes of place, comparatively few persons
ever see, can perhaps hardly be called an interesting object
for an opera-glass, and yet the beauty of the planet is greatly
increased when viewed with such aid.   Mercury is brilliant
enough to be readily distinguishable, even while the twilight
is still pretty bright ; and I have had most charming views of
the shy planet, glittering like a globule of shining metal
through the fading curtain of a winter sunset.

Venus is, under favorable circumstances, a very interesting
planet for opera-glass observations.   The crescent phase can
be seen with a powerful glass near inferior conjunction, and,
even when the form of the planet can not be discerned, its exceeding brilliancy makes it an attractive object.   The flood of
light which Venus pours forth, and which is so dazzling
that it baffles the best telescopes, to a greater or less extent,
in any effort to descry the features of that resplendent disk,
is evidently reflected from a cloud-burdened atmosphere.
While these clouds render the planet surprisingly lustrous
to our eyes, they must, of course, keep the globe beneath them most of the time in shadow.   It is a source of
keen regret that the surface of Venus can not be seen as
clearly as that of Mars, for, *a priori*, there is rather more
reason to regard Venus as possibly an inhabited world than
any other of the Earth's sister planets, not excepting Mars.
Still, even if we could plainly make out the presence of
oceans and continents on Venus, that fact would hardly be
any better indication of the possibility of life there than is
furnished by the phenomena of its atmosphere.   It is an
interesting reflection that in admiring the brilliancy of this
10

splendid planet the light that produces so striking an effect upon our eyes has but a few minutes before traversed the atmosphere of a distant world, which, like our own air, may furnish the breath of life to millions of intelligent creatures, and vibrate with the music of tongues speaking languages as expressive as those of the earth.

Mars, being both more distant and smaller than Venus, does not present so splendid a scene, and yet when it is at or near opposition it is a superb object even for an opera-glass, its deep reddish-yellow color presenting a fine contrast to that of most of the stars. It can often be seen in conjunction with, or near to, the moon and stars, and the beauty of these phenomena is in some cases greatly enhanced by the use of a glass. To find Mars (and the same remark applies to the other planets), take its right ascension and declination for the required date from the Nautical Almanac, and then mark its place upon a planisphere or any good star-map. This planet is at the present time (1888) slowly drawing nearer to the earth at each opposition, and in 1892 it will be closer to us than at any time since 1877, when its two minute satellites were discovered. It will consequently grow brighter every year until then. How splendidly it shines when at its nearest approach to the earth may be inferred from the fact that in 1719 it was so brilliant as actually to cause a panic. This was doubtless owing to its peculiar redness. I well remember the almost startling appearance which the planet presented in the autumn of 1877. Mars is especially interesting because of the apparently growing belief that it may be an inhabited world, and because of certain curious markings on its surface that can only be seen under favorable conditions. The recent completion of the great Lick telescope and other large glasses, and the approach of the planet to a favorable opposition, give reason to hope that within the next few years a great deal of light will be cast upon some of the enigmatical features of Mars's surface.

Jupiter, although much more distant than Mars, is ordinarily a far more conspicuous phenomenon in the sky on account of his vast bulk. His interest to observers with an opera-glass depends mainly upon his four moons, which, as they circle about him, present a miniature of the solar system. With a strong opera-glass one or two of Jupiter's little family of moons may occasionally be caught sight of as excessively minute dots of light half-hidden in the glare of the planet. If you succeed under favorable circumstances in seeing one of these moons with your glass, you will be all the more astonished to learn that there are several apparently well-authenticated instances of one of the moons of Jupiter having been seen with the naked eye.

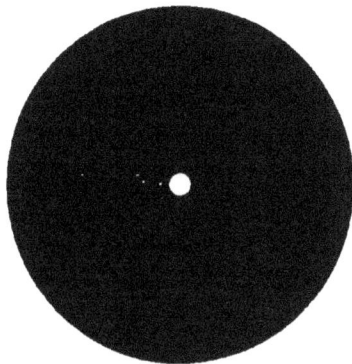

JUPITER AND HIS MOONS. (SEEN WITH A FIELD-GLASS ; SEVEN DIAMETERS.)

With a field-glass, however, you will have no difficulty in seeing all of the moons when they are properly situated. If you miss one or more of them, you may know that it is either between you and the planet, or behind the planet, or buried in the planet's shadow, or else so close to the planet as to be concealed by its radiance.

It will be best for the observer to take out of the Nautical Almanac the "configurations of Jupiter's satellites" for the evenings on which he intends to make his observations, recollecting that the position of the whole system, as there given, is reversed, or presented as seen with an astronomical telescope, which inverts objects looked at, as an opera-glass does not. In order to bring the satellites into the positions in which he will see them, our observer has only to turn the page in the Nautical Almanac showing their configurations upside down.

Of course, since the motions of the satellites, particularly of the inner ones, are very rapid, their positions are continually changing, and their configurations are different every night. If the observer has any doubt about his identification of them, or thinks they may be little stars, he has only to carefully note their position and then look at them again the next evening. He may even notice their motion in the course of a single evening, if he begins early and follows them for three or four hours. It is impossible to describe the peculiar attractions of the scene presented by the great planet and his four little moons on a serene evening to an observer armed with a powerful glass. Probably much of the impressiveness of the spectacle is owing to the knowledge that those little points of light, shining now in a row and now in a cluster, are actually, at every instant, under the government of their giant neighbor and master, and that as we look upon them, obediently making their circuits about him, never venturing beyond a certain distance away, we behold a type of that gravitational mastery to which our own little planet is subject as it revolves around its still greater ruler, the sun, to whose control even Jupiter in his turn must submit.

The beautiful planet Saturn requires for the observation of its rings magnifying powers far beyond those of the instruments with which our readers are supposed to be armed. It would be well, however, for the observer to trace its slow motion among the stars with the aid of the Nautical Almanac, and he should be able with a good field-glass to see, under favorable circumstances, the largest of its eight moons, Titan. This is equal in brilliancy to an 8·5 magnitude star. Its position with respect to Saturn on any given date can be learned from the Ephemeris.

It may appear somewhat presumptuous to place Uranus, a planet which it required the telescope and the eye of a Herschel to discover, in a list of objects for the opera-glass. But it must not be forgotten that Uranus was seen certainly several, and probably many, times before Herschel's discovery, being simply mistaken, on account of the slowness of its motion, for a fixed star. When near opposition, Uranus looks as bright as a sixth-magnitude star, and can be easily detected with the naked eye when its position is known. With an opera-glass (and still more readily with a field-glass) this distant planet can be watched as it moves deliberately onward in its gigantic orbit. Its passage by neighboring stars is an exceedingly interesting phenomenon, and it is in this way that you may recognize the planet.

On the evening of May 29, 1888, I knew, from the co-ordinates given in the Nautical Almanac, that Uranus was to be found a short distance east of Mars, which was then only a few degrees from the well-known star Gamma Virginis. Accordingly, I turned my opera-glass upon Mars, and at once saw a star in the expected position, which I knew was Uranus. But there were other small stars in the field, and, supposing I had not been certain which was Uranus, how could I have recognized it? The answer is plain : simply by watching for a night or two to see which star moved. That star would, of course, be Uranus. The accompanying cuts will show the motions of Mars and Uranus with respect to neighboring stars at that time, and will serve as an example of the method of

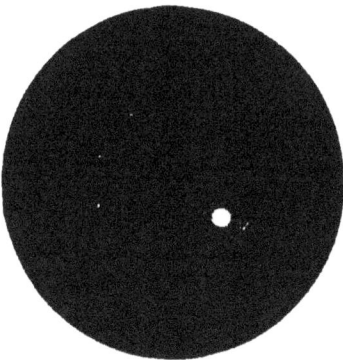

MARS AND URANUS, MAY 29, 1888.   MARS AND URANUS, JUNE 1, 1888.

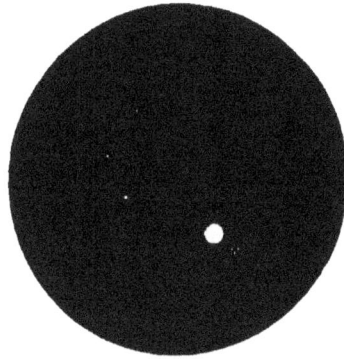

distinguishing a planet from the fixed stars by its change of place. In the first cut we have the two planets and three neighboring stars as they appeared on May 29th. These stars were best seen with a field-glass, although an opera-glass readily showed them.

On June 1st the relative positions of the planets and stars were as shown in the second cut. A glance suffices to show that not only Mars but Uranus also has shifted its position with respect to the three immovable stars. This change of place alone would have sufficed to indicate the identity of Uranus. To make sure, the inexperienced observer had only to continue his observations a few nights longer.

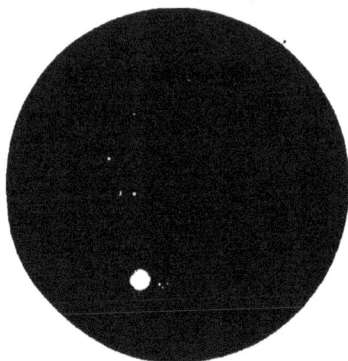

MARS AND URANUS, JUNE 6, 1888.

On June 6th Mars and Uranus were in conjunction, and their position, as well as that of the same set of three stars, is shown in the third cut. It will be seen that while Mars had changed its place very much more than Uranus, yet that the latter planet had now moved so far from its original position on May 29th, that there could be no possibility that the merest tyro in star-gazing would fail to notice the change. Whenever the observer sees an object which he suspects to be a planet, he can satisfy himself of its identity by making a series of little sketches like the above, showing the position of the suspected object on successive evenings, with respect to neighboring stars. The same plan suffices to identify the larger planets, in the case of which no glass is necessary. The observer can simply make a careful estimate by the naked eye of the supposed planet's distance and bearing from large stars near it, and compare them with similar observations made on subsequent evenings.

THE SUN.—That spots upon the sun may be seen with no greater optical aid than that of an opera-glass is perhaps well known to many of my readers, for during the past ten years public attention has been drawn to sun-spots in an especial manner, on account of their supposed connection with meteorology, and in that time there have been many spots upon the solar disk which could not only be seen with an opera-glass, but even with the unassisted eye. At present (1888) we are near a minimum period of sun-spots, and the number to be seen even with a telescope is comparatively very small, yet only a few days before this page was written there was a spot on the sun large enough to be conspicuous with the aid of a field-glass. During the time of a spot-maximum the sun is occasionally a wonderful object, no matter how small the power of the in-strument used in viewing it may be. Strings of spots of every variety of shape sometimes extend completely across the disk. Our illustration shows the appear-ance of the sun, as drawn by the au-thor on the 1st of September, 1883. Every one of the spots and spot-groups there repre-sented could be seen with a good field-glass, and nearly all of them with an opera-glass.

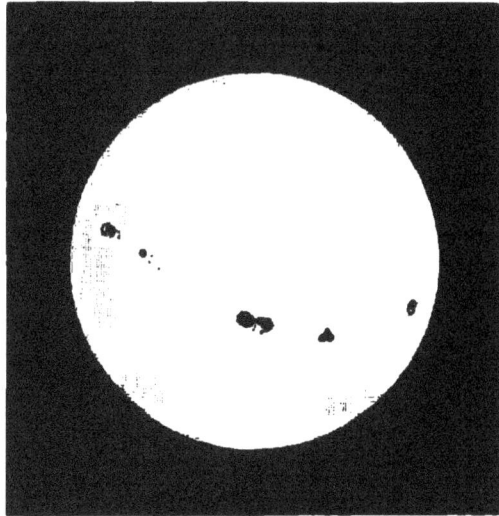

THE SUN, SEPTEMBER 1, 1883.

As in all such cases, our interest in the phenomena in-creases in proportion to our understanding of their significance and their true scale of magnitude. In glancing from side to side of the sun's disk, the eye ranges over a distance of more than 860,000 miles—not a mere ideal distance, or an expanse of empty space, but a distance filled by an actual and, so to speak, tangible body, whose diameter is of that stupendous magnitude. One sees at a glance, then, the enormous scale on which these spots are formed. The earth placed beside them would be but a speck, and yet they are mere pits in the surface of the sun, filled perhaps with partially cooled metallic vapors, which have been cast up from the interior, and are settling back again. It is worth anybody's while to get a glimpse at a sun-spot if he can, for, although he may see it merely as a black dot on the shining disk, yet it represents the play of physical forces whose might and power are there exercised on a scale really beyond human comprehension. The imagination of Milton or Dante would have beheld the mouth of hell yawning in a sun-spot.

In order to view the sun it is, of course, necessary to contrive some protection for the eyes. This may be constructed by taking two strips of glass four or five inches long and an inch wide, and smoking one of them until you can without discomfort look at the sun through it. Then place the two strips together, with the smoked surface inside—taking care to separate them slightly by pieces of cardboard placed between the ends—and fasten the edges together with strips of paper gummed on. Then, by means of a rubber band, fasten the dark glass thus prepared over the eye-end of your opera-glass in such a way that both of the lenses are completely covered by it. It will require a little practice to enable you to get the sun into the field of view and keep it there, and for this purpose you should assume a posture— sitting, if possible—which will enable you to hold the glass very steady. Then point the glass nearly in the direction of the sun, and move it slowly about until the disk comes in sight. It is best to carefully focus your instrument on some distant object before trying to look at the sun with it. As there is some danger of the shade-glass being cracked by the heat, especially if the object-glasses of the instrument are pretty large, it would be well to get the strips of glass for the shade large enough to cover the object-end of the instrument instead of the eye-end. At a little expense an optician will furnish you with strips of glass of complementary tints, which, when fastened together, give a very pleasing view of the sun without discoloring the disk. Dark red with dark blue or green answer very well; but the color must be very deep. The same arrangement, of course, will serve for viewing an eclipse of the sun.

A word, finally, about the messenger which brings to us all the knowledge we possess of the contents and marvels of space—light. Without the all-pervading luminiferous ether, narrow indeed would be our acquaintance with the physical creation. This is a sympathetic bond by which we may conceive that intelligent creatures throughout the universe are united. Light tells us of the existence of suns and systems so remote that the mind shrinks from the attempt to conceive their distance; and light bears back again to them a similar message in the feeble glimmering of our own sun. And can any one believe that there are no eyes out yonder to receive, and no intelligence to interpret that message?

Sir Humphry Davy has beautifully expressed a similar thought in one of his philosophical romances:

In Jupiter you would see creatures similar to those in Saturn, but with different powers of locomotion; in Mars and Venus you would find races of created forms more analogous to those belonging to the Earth; but in every part of the planetary system you would find one character peculiar to all intelligent natures, a sense of receiving impressions from light by various organs of vision, and toward this result you can not but perceive that all the arrangements and motions of the planetary bodies, their satellites and atmospheres, are subservient. The spiritual natures, therefore, that pass from system to system in progression toward power and knowledge preserve at least this one invariable character, and their intellectual life may be said to depend more or less upon the influence of light.*

Light is a result, and an expression, of the energy of cosmical life. The universe lives while light exists. But when the throbbing energies of all the suns are exhausted, and space is filled with universal gloom, the light of intelligence must vanish too.

One can not read the wonderful messages of light—one can not study the sun, the moon, and the stars in any manner —without perceiving that the physical universe is enormously greater than he had thought, and that the creation, of which the Earth is an infinitesimal part, is almost infinitely more magnificent in actual magnitude than the imaginary domain which men of old times pictured as the dwelling-place of the all-controlling gods ; without feeling that he has risen to a higher plane, and that his intellectual life has taken a nobler aim and a broader scope.

* See " Consolations in Travel, or the Last Days of a Philosopher "; Dialogue I.

INDEX.

THE END.